S0-BSB-471

WITHDRAWN

Georg Wüst

Studies in Physical Oceanography

Volume 1

Studies in Physical Oceanography

A Tribute to Georg Wüst
on his 80th Birthday

Volume 1

Edited by

ARNOLD L. GORDON

Lamont-Doherty Geological Observatory
of Columbia University
Palisades, New York 10964

GORDON AND BREACH SCIENCE PUBLISHERS
New York　　　　　London　　　　　Paris

Copyright © 1972 by

Gordon and Breach, Science Publishers, Inc.
440 Park Avenue South
New York, N. Y. 10016

Editorial office for the United Kingdom

Gordon and Breach, Science Publishers Ltd.
42 William IV Street
London W. C. 2

Editorial office for France

Gordon & Breach
7–9 rue Emile Dubois
Paris 14^e

551.46
St9
84891
Sept/1923

Library of Congress catalog card number 75–162627. ISBN 0 677 15160 8. All rights reserved. No part of this book may be reproduced or utilized in any form or by any means, electronic or mechanical, including photocopying, recording, or by any information storage and retrieval system, without permission in writing from the publishers. Printed in east Germany.

Editor's Preface

The collection of articles composing this volume, represent the high esteem in which Georg Wüst is held by his colleagues. In each of the fields discussed in these studies Wüst has made important contributions; often his own articles are now considered the classical studies on which the present work is based. The contributions are ordered to match the seven major fields of study of Wüst, as given in the dedication by Guenter Dietrich.

In preparing this volume I unfortunately got off to a late start, so that the authors hadn't much time to prepare their contributions. This regrettably made it impossible for many of the "friends of Wüst" to contribute. There is also the distinct possibility I overlooked, due to ignorance, some oceanographers who would have liked to share in this tribute. This results from my contact with Wüst being limited to the last decade, while his activity in oceanography has spanned the last five and one half decades.

I attempted to limit the list of authors to those who know Wüst personally. Had I invited all oceanographers who are working in fields where Wüst has made important contributions to join in this "Tribute to Georg Wüst on His 80th Birthday" it would have filled many volumes and been far beyond my means to edit the collection.

Georg Wüst celebrated his 80th birthday on 15 June 1970. At that time copies of the dedication and most of the papers were available and sent to Wüst at his home in Erlangen, Germany. Thus the main purpose of this volume, i.e. to present to Wüst on his birthday a collection of papers dedicated to him in friendship, was achieved nearly in total. However it will be near his 81st birthday that all the papers will be published for general distribution.

My contact with Wüst dates from 1961 when I began my graduate studies at Columbia University and Wüst was visiting professor. He remained in this capacity until 1964, during which time I worked closely with him on his Caribbean Sea studies. I learned from him that through combination of data and physical-mathematical analysis with imagination one can gain fairly detailed understanding of the ocean structure and circulation. The Wüst technique used with the computer and present day mathematical methods

opens the door to a field which can be called modern descriptive oceanography. The results of this field form the basis on which the theoretical
oceanographers must build.

Acknowledgements: In editing this volume I relied on the aid of Dr. Ste-
phen Eittreim of Lamont-Doherty Geological Observatory, who looked
after sending the contributions out for review during the summer of 1970
when I was at sea. His aid is much appreciated; it permitted progress to be
made during the summer. Thanks are also due to the many oceanographers
who reviewed the manuscripts. Their suggestions were most valuable in
weeding out the minor inconsistencies or improper wording which so often
creep into manuscripts and escape the view of the authors who are so close
to their work. The secretarial work of Mrs. Jeanne Stolz (who also was
Prof. Wüst's secretary during his days at Columbia University) is much appreciated. I also thank my wife, Susan, for her understanding during the
times I spoke of nothing but the "Wüst Volume".

Special thanks go to the director of Lamont-Doherty Geological Observatory, Dr. Maurice Ewing, who was responsible for bringing Prof. Georg
Wüst to Columbia, hence making it possible for direct contact of the scientists and students at Columbia with Georg Wüst.

ARNOLD L. GORDON

Palisades, New York

Table of Contents

VOLUME 2

Georg Wüst's Scientific Work
Dedication to his Eightieth Birthday

It is unusual for a scientist to have a list of publications covering a period of half a century. But such is true for Georg Wüst, who celebrates his eightieth birthday on the 15th of June, 1970. His list bears evidence to the work of a single person which has benefited the whole scientific world of marine sciences. This work is not taken as a matter of course, but is acknowledged with sincere thanks.

Seven main topics cut like a red file through Wüst's long publications list. He went over these topics repeatedly, giving new insights and still valid results. He wrote without padding. Characteristic of him is that he gave many of his results in informative figures, ready for publication in textbooks to which authors like to refer.

The seven main topics on which Wüst has worked can be outlined as follows:

1. Evaporation and water budget of the world ocean.
2. Vertical circulation of the Atlantic Ocean as the central topic of the "Meteor" Expedition, 1925–27.
3. Geostrophic movement in the Gulf Stream and Kuroshio systems.
4. Bottom-circulation in the world ocean and its dependence, as deduced from the topographical main features of the world ocean.
5. Atlantic deep-sea circulation.
6. Circulation of the Mediterranean Seas.
7. The history of deep-sea research.

Wüst started his scientific career in 1914 with his doctoral thesis on evaporation (Lit. No.1)*, which was published in 1920. He developed applicable reduction methods for evaporation measurements after he had found how dubious was the knowledge about evaporation. Through his work, the water budget of the sea came within reach of scientific investigation

---

* Literature numbers refer to G. Wüst's list of publications at end of dedication.

ard the relationships between evaporation—rainfall—run off ard salinity content of the ocean surface were brought to quantitative focus. His methods ard results are described in publications 1, 3, 42, 46, 48, 55, 77, 79, 81, and 82.

The second main topic, vertical circulation in the Atlantic Ocean, was the central problem of the "Deutsche Atlantische Expedition" with the R.V. "Meteor" 1925–27, which carried out 14 cross sections (310 stations) between 20° N ard 64° S. Wüst formulated the topic of this large maritime undertaking together with his teacher Alfred Merz, who was the initiator and organizer. Publications 2, 6, 7, and 9 refer to it. Because of serious illness, Merz left the ship after the fifth station in Buenos Aires. He died there two months later. According to his wishes, Wüst took over the guidance of the oceanographic work.

Closely related to the central problem of the expedition is a third group of works that treated the geostrophic currents in the sea. The first investigation (Lit. No. 8) has become a classic. Using the Florida Current as an example Wüst confirmed the reliability of the current velocities observed by current meters and calculated from the pressure field. Therefore, this work has given confidence to the oceanographers that the "dynamic method" leads to the real stationary current field. At the same time there was included in this work of the "Meteor" Expedition, the future task of evaluating the total circulation from the direct current measurements of 10 anchor stations and from the observed density stratifications on 14 transatlantic profiles. The anchor station did not preduce useful results of the circulation because of the swing movements of the ship as well as those of the tidal currents. A. Defant pointed this out, but it was up to Wüst to evaluate the geostrophic currents: see 84, 87, 89. The work on the Florida Current contributed to the fact that his interest in the Gulf Stream was aroused and maintained: see 14, 27, 41, 50, 52, 63, 64, 65; and that he was inspired to comparisons with its Pacific counterpart, the Kurcshio: 41, 50.

The fourth main topic concerns the representation of the circulation of the bottom water and also the large-scale structure of the deep-sea bottom which explains the spreading of the bottom water. The Weddell Sea and the Greenland Sea were fourd to be areas of origin of the bottom-water circulation in the three oceans. The distribution of the bottom water is explained in detail ard the large-scale structure in 60 basins is traced in the following: 19, 20, 21, 22, 26, 29, 30, 40, 47, 51, 54, 59, 60, 62, 69, and 70. Wüsts's interpretations of the large-scale structure of the world ocean has remained valid up to now, although it does not correspond with the new classification of the physiographic provinces by Heezen et al. (1959), which takes the origin of the oceans into consideration; but in the case of the distribution of the bottom water, topography is decisive.

Topic five, the Atlantic deep-sea circulation, its layering and movement, became the important part of the treatment of the "Meteor" Expedition. Wüst made comparisons with the other oceans: 12, 15, 28. After that he worked on the Atlantic stratosphere: see 10, 31, 33, 34, 38, 39, 44, 53, 57, 72, 73, 74, 76, 78, and 109. Of these publications, No. 31 is the most important and the most extensive. It has contributed to the fact that, for solidity, the publication of the results of the "Meteor" expedition has not been matched by any other comparable publication. Together with the contributions of A. Defant, investigator of the oceanic troposphere, a high standard was reached concerning the exploration of a whole ocean area.

The circulation of the Mediterranean Seas was the sixth main topic pursued with great intensity by Wüst, who was especially concerned with four seas: the North Polar Sea: 66, 67, 68; the Baltic Sea: 75, 83, 88, 95; the European Mediterranean Sea: 91, 92, 94, 97, 99; and the Caribbean Sea: 100, 105, 106. The core-layer method developed and applied by Wüst, as well as the critical treatment of the extensive observation material, has enhanced our knowledge of the circulation of each of these seas. This work provides a base on which open questions can be met consequently in future, and this is the real value of Wüst's efforts. In much as the Atlantic Program in the International Geophysical Year 1957/58 has been founded on the "Meteor" Expedition 1925/27, future undertakings in the Mediterranean Seas will find Wüst's results of great help.

Any scientist who has helped in the exploration of the deep oceans seas for half a century knows the subject thoroughly and can advise the following generations from his own experience. Wüst does so under the seventh topic, the history of deep-sea exploration: 85, 90, 101, 102, 110. It is through his detailed studies, with regard to layering and circulation of oceans and Mediterranean Seas, which enabled him to write such historic contributions. He described the limitations of the classical works which were based on single observations with Nansen bottles and reversing thermometers, with the assumption of stationary conditions in the deep-sea. The classical works deal with the macro-structure of layering, circulation, and geostrophic currents. These results do have lasting value because they delineate the general ocean structure and circulation and aid in the preparation of special studies concerning micro structure and time variability of the sea. Only now it is possible, because of the electronic approach to deep-sea exploration with continuous recordings, to treat these problems as has been done since 1964 on the cruises of the new research vessels.

Besides Wüst's admirable research his activities in Berlin at the Institut für Meereskunde until 1945, as the Director of the Institut für Meereskunde in Kiel from 1946 to 1959, and as a visiting professor at the Lamont-Doherty

Geological Observatory of Columbia University during 1960–1964 are distinctive. He deserves recognition for all of the above, and also for his continuous contribution to reestablish the institute in Kiel.

Wüst is a Berliner, frank and involved, amiable and temperately judging, busy and tough, flexible and steady. All these properties make a man win friends as well as opponents. But a man who has dedicated his life to marine research can be sure of the high esteem of his colleagues, pupils, and friends.

PROF. GUENTER DIETRICH

Institut für Meereskunde
an der Universität Kiel
Kiel, W. Germany

List of scientific publications of Georg Wüst

1. Die Verdunstung auf dem Meere. (Veröffentl. Inst. f. Meereskunde, N. F. Reihe A, H. 6, Berlin 1920, S. 1–96.)
2. (Gemeinsam mit *A. Merz*:) Die atlantische Vertikalzirkulation (Zeitschr. Ges. f. Erdkunde Berlin 1922, S. 1–35.)
3. Verdunstung und Niederschlag auf der Erde. (Zeitschr. Ges. f. Erdkunde, Berlin 1922, S. 33–43.)
4. Ältere und neuere Anschauungen über die Strömungen der Nordsee. (Die Naturwissenschaften, Berlin 1923, H. 11, S. 199–202.)
5. Die ersten akustischen Tiefseelotungen. (Die Naturwissenschaften, Berlin 1923, H. 15, S. 282–88.)
6. (Gemeinsam mit *A. Merz*:) Die atlantische Vertikalzirkulation. Dritter Beitrag. (Zeitschr. Ges. f. Erdkunde, Berlin 1923, S. 132–44.)
7. (Gemeinsam mit *A. Merz*:) Die atlantische Vertikalzirkulation. Eine Erwiderung an Prof. *O. Pettersson*. (Ann. d. Hydr. 1923, S. 149f.)
8. Florida- und Antillenstrom. Eine hydrodynamische Untersuchung. (Veröff. Inst. f. Meereskunde, N. F., Reihe A, H. 12, Berlin 1924, S. 1–48.)
9. Die Deutsche Atlantische Expedition auf dem Forschungsschiff "Meteor". Berichte über die ozeanographischen Untersuchungen I–IV. (Zeitschr. Ges. f. Erdkunde, Berlin 1926/27. 71 S.)
10. Der Ursprung der Atlantischen Tiefenwässer. (Jubiläums-Sonderband Zeitschr. Ges. f. Erdkunde, Berlin 1928, S. 507–34.)
11. Ozeanographische Methoden und Instrumente der Deutschen Atlantischen Expedition (Erg.-H. III, Zeitschr. Ges. f. Erdkunde, Berlin 1928, S. 66–83.)
12. Schichtung und Tiefenzirkulation des Pazifischen Ozeans auf Grund zweier Längsschnitte. (Veröff. Inst. f. Meereskunde, N.F., Reihe A, H. 20, Berlin 1929. S. 1–64.) Mit 14 Abb. und 4 Tafeln.
13. Das Bouvet-Problem. (Zeitschr. Ges. f. Erdkunde, Berlin 1929, S. 133–42.)
14. Der Golfstrom. (Zeitschr. Ges. f. Erdkunde, Berlin 1930, H. 1/2, S. 42–49.)
15. Meridionale Schichtung und Tiefenzirkulation in den Westhälften der drei Ozeane. (Journal du Conseil, Kopenhagen 1930, Vol. V, Nr. 1, S. 7–21.)
16. (Gemeinsam mit *A. Defant*:) Die Mischung von Wasserkörpern im System $S = f(t)$. (Rapp. et procès-verbaux, Vol. LXVII, 1930, S. 40–47.)
17. Ozeanographische Instrumente und Methoden. (Wiss. Ergebn. d. D. A. E. auf Verm.- und Forschungsschiff "Meteor" 1925–1927, Bd. IV, I. Teil, Berlin 1932, S. 1–178.)
18. Das ozeanographische Beobachtungsmaterial (Serienmessungen). (Wiss. Erg. d. D. A. E. auf Verm.- u. Forschungsschiff "Meteor" 1925–1927, Bd. IV, II. Teil, Berlin 1932, S. 1–290.)
19. Bodenwasser und Bodenkonfiguration der atlantischen Tiefsee. (Zeitschr. Ges. f. Erdkunde, Berlin 1933, Nr. 1/2, S. 1–18.)
20. Das Bodenwasser und die Gliederung der atlantischen Tiefsee. Wiss. Erg. d. D. A. E. "Meteor" 1925–1927, Bd. VI, 1. Lfg., Berlin 1933, S. 1–107.)

21. Bodenwasser und Bodenkonfiguration der atlantischen Tiefsee. (Forschungen und Fortschritte, 9. Jahrg., Berlin 1933, Nr. 7, S. 96f.)

22. Anzeichen von Beziehungen zwischen Bodenstrom und Relief in der Tiefsee des Indischen Ozeans. (Naturwissenschaften, 22, H. 16, Berlin 1934, S. 241–44.)

23. Salzgehalt und Wasserbewegung im Suezkanal. (Naturwissenschaften, 22, H. 26, Berlin 1934, S. 446–50.)

24. Neue Anschauungen über Eisberge. (Peterm. Mitt., 1934, H. 6, S. 176.)

25. Thermometric measurement of depth. (The Hydrographic Review, Vol. X, 2, S. 28–49, Monaco 1933.)

26. Über die Bedeutung von Bodentemperaturmessungen für die ozeanographische, morphologische, chemische und geologische Erforschung der Tiefsee. (James Johnstone Memorial Volume, Liverpool 1934, S. 242–56.)

27. Das Golfstromproblem. In: Tiefseebuch (Band III der Sammlung "Das Meer"), Berlin 1934, S. 125–42.

28. Zur Frage des indischen Tiefseestroms. (Naturwissenschaften, 22, H. 9, Berlin 1935, S. 137–39.)

29. Die Ausbreitung des antarktischen Bodenwassers im Atlantischen und Indischen Ozean. (Zeitschr. f. Geophysik, 11, H. 1/2, Braunschweig 1935, S. 40–49.)

30. (Zusammen mit *Th. Stocks*:) Die Tiefenverhältnisse des offenen Atlantischen Ozeans. (Begleitworte zur Übersichtskarte 1 : 20 Mill.) (Wiss. Erg. d. D. A. E. "Meteor" 1925–1927, Bd. III, I. Teil, Berlin 1935, S. 1–32.)

31. Die Stratosphäre des Atlantischen Ozeans. (Wiss. Erg. d. D. A. E. "Meteor" 1925–1927, Bd. VI, 2. Lfg., Berlin 1935, S. 1–288. Mit Atlas.)

32. Fortschreitende Salzgehaltsabnahme in Suezkanal. (Ann. d. Hydr., Berlin 1935, S. 391–95.)

33. Die Vertikalschnitte der Temperatur, des Salzgehaltes und der Dichte. (Teil A des "Atlas" zu Bd. VI der Wiss. Erg. d. D. A. E. "Meteor" 1925–27, Berlin 1936, Beilage II–XLVI.

34. Die Horizontalkarten der Temperatur, des Salzgehaltes und der Dichte. (Ebenda, Beilage XLVII–XCI.)

35. Erster Versuch einer synthetisch-regionalen Behandlung des Weltmeeres. (Peterm. Mitt., Gotha 1935, S. 410.)

36. Humboldtstrom, nicht Perustrom. (Peterm. Mitt., 1935, H. 12, S. 439–41.)

37. Die Gliederung des Weltmeeres. Versuch einer systematischen ozeanographischen Namengebung. (Peterm. Mitt., 1936, S. 33–38. Mit 1 Tab. u. 2 Taf.)

38. Die Tiefenzirkulation im Raume des Atlantischen Ozeans. (Naturwissenschaften, 24, Berlin 1936, H. 9, S. 133–41.)

39. Deep circulation in the expanse of the North Atlantic Ocean. (The Hydrographic Review, Vol. XXII, 2, Monaco 1936, S. 23–31.)

40. Die Gliederung des Weltmeeres. (The divisions of the oceans.) (The Hydrographic Review, Vol. XIII, 2, Monaco 1936, S. 36–54.)

41. Kuroshio und Golfstrom. Eine vergleichende hydrodynamische Untersuchung. (Veröff. Inst. f. Meereskunde, N. F., Reihe A, H. 29, Berlin 1936, S. 1–69.)

42. Oberflächensalzgehalt, Verdunstung und Niederschlag auf dem Weltmeere. (Festschr. N. Krebs, Stuttgart 1936, S. 347–59.)

43. Die Erforschung der Bodenkonfiguration des Austral-asiatischen Mittelmeeres durch die holländische "Snellius"-Expedition. (Zeitschr. Ges. für Erdkunde, Berlin 1936, S. 347–58.)

44. Die Stratosphäre des Atlantischen Ozeans. (Zeitschr. für Geophysik, Braunschweig 1936, S. 287.)

45. Zur Frage der Eintragung von Lotzahlen in die amtlichen Seekarten und in wissenschaftliche Tiefenkarten. (Zeitschr. Ges. f. Erdkunde, Berlin 1937, H. 1/2, S. 54 f.)

46. Temperatur- und Dampfdruckgefälle in den untersten Metern über der Meeresoberfläche. (Met. Zeitschr., Braunschweig 1937, S. 4–9.)

47. Die Großgliederung des Tiefseebodens, zugleich ein Vorschlag einer systematischen geographischen Namengebung für die Tiefseebecken des Weltmeeres. (Ass. Océanogr. Phys. Procès-Verb. Nr. 2, 1937, S. 70–72.)

48. Oberflächensalzgehalt, Verdunstung und Niederschlag auf dem Weltmeere. (Procès-Verb. Nr. 2, 1937, Berlin, S. 79–83.)

49. Zur Frage: Perustrom oder Humboldtstrom. Eine Erwiderung auf G. Schotts gleichnamigen Aufsatz. (Ann. d. Hydr., April 1937, Berlin, S. 172–74.)

50. Golfstrom u. Kuroshio in ihrer Beziehung für das Klima der nördlichen Halbkugel. (Geistige Arbeit, Berlin 1937, 4. Jahrg., Nr. 13, S. 5 f.)

51. Bodentemperatur und Bodenstrom in der pazifischen Tiefsee. (Veröff. Inst. f. Meereskunde, N. F., Reihe A, H. 35, Berlin 1937, S. 1–56.)

52. Neuere Auffassungen über das Wesen des Golfstromsystems und die Benennung seiner Glieder. (Der Seewart, H. 10/11, Hamburg 1937, S. 359–67, Taf. 3.)

53. La circulation générale dans les océans. (Compt. Rend. Congrès Intern. Géographie, Amsterdam 1938. Rapports, Supplément, Leiden 1938, S. 3.)

54. Das Problem des antarktischen Bodenstroms im Weltmeer. (Ebenda, Tome II, Sect. II b: Ozeanographie, Leiden 1938, S. 45–48.)

55. Niederschlags- und Verdunstungsmessungen auf der Ostsee. (VI. Balt. Hydrolog. Konferenz, Anf. 1938, Hauptbericht 8).

56. Der Internationale Geographenkongreß in Amsterdam. Sect. II b: Ozeanographie. (Peterm. Mitt., 1938, H. 9, S. 272 f.)

57. Die dynamischen Werte für die Standardtiefen an den Beobachtungsstationen. (Unter Mitarbeit von A. Defant.) (Wiss. Erg. d. D. A. E. "Meteor" 1925–27, Bd. VI, II. Teil, Berlin 1938, S. 99–180.)

58. Vier Strömungskarten des Südpolarmeeres. Mittlere Wasserbewegung an der Oberfläche im Südsommer. (Beiblätter z. D. Adm.-Karten 1061–1064.)

59. Bodentemperatur und Bodenstrom in der atlantischen, indischen und pazifischen Tiefsee. (Gerlands Beiträge zur Geophysik, Bd. 54, 1, 1938, S. 1–8.)

60. Die Großgliederung des atlantischen Tiefseebodens. (Geologische Rundschau, Bd. XXX, 1938, H. 1/2, Stuttgart, S. 132–37.)

61. Das "Snellius"-Werk (Vol. I und II). (Zeitschr. d. Ges. f. Erdkunde, Berlin 1939, S. 102–08.)

62. Die Grenzen der Ozeane und ihrer Nebenmeere. (Beiheft zum Mai-Heft 1939 der "Ann. d. Hydr. u. Marit. Met.".)

63. Das submarine Relief bei den Azoren. (Sonderausgabe aus den Abhandlungen d. Preuß. Akad. d. Wiss., Jahrg. 1939, Phys.-Math. Klasse Nr. 5, Berlin 1939, S. 46–58.)

64. Das Relief des Azorensockels und des Meeresbodens nordwestlich der Azoren. (August-Beiheft zu den "Ann. d. Hydr. usw.", 1940, S. 1–19; nit 7 Abb. u. 5 Taf.)

65. Die auf den Stationen des Forschungsschiffes "Altair" ausgeführten ozeanographischen Reihenmessungen. (März-Beiheft zu den "Ann. d. Hydr.", 1941, S. 1–60; mit 5 Abb.)

66. Relief und Bodenwasser im Nordpolarbecken. (Zeitschr. Ges. f. Erdkunde, Berlin 1941, S. 163–80.)

67. Die morphologischen und ozeanographischen Verhältnisse des Nordpolarbeckens. (Veröff. d. Dt. Wiss. Inst. in Kopenhagen, Reihe I, Arktis, H. Nr. 6, Berlin 1942, S. 1–21; mit 1 Karte.)
68. Neuere Anschauungen über die Morphologie und Ozeanographie des Nordpolarbeckens. (Forschungen u. Fortschritte, 1943, Nr. 9/10, S. 110–13.)
69. Die geographische Einteilung und Benennung des Weltmeeres. (Marine-Rundschau, 48, Jg. 1943, H. 3, S. 156.)
70. Der subarktische Bodenstrom in der westatlantischen Mulde. (Ann. d. Hydr. usw., 1943, S. 249–55; mit 2 Taf.)
71. Wiederauffindung eines submarinen, vorübergehend inselbildenden Vulkans bei San Miguel (Azoren). (Zeitschr. d. Ges. f. Erdkunde, Berlin 1944, Heft 3/4, S. 85–92.)
72. Die Temperaturinversion im Tiefenwasser des Südatlantischen Ozeans. (Deutsche Hydrogr. Ztschr., 1948, S. 109–24.)
73. Über die Zweiteilung der Hydrosphäre. (Deutsche Hydrogr. Zeitschr., 1949, S. 218–225.)
74. Die Kreisläufe der atlantischen Wassermassen, ein neuer Versuch räumlicher Darstellung. (Forschungen u. Fortschritte, 1949, S. 287–90.)
75. Ergänzende Bemerkungen zu H. Wattenbergs Aufsatz über die natürliche Einteilung der Ostsee. (Kieler Meeresforschungen, Bd. VI, 1949, S. 15–17.)
76. Blockdiagramme der atlantischen Zirkulation auf Grund der "Meteor"-Ergebnisse. (Kieler Meeresforschungen, Bd. VII, 1950, S. 24–34.)
77. Wasserdampf und Niederschlag auf dem Meere als Glieder des Wasserkreislaufs (unter besonderer Berücksichtigung von Ergebnissen der "Meteor"-Expedition und neuerer Arbeiten). (Deutsche Hydrographische Zeitschrift, 1950, S. 111–27.)
78. Über die Fernwirkungen antarktischer und nordatlantischer Wassermassen in den Tiefen des Weltmeeres. (Naturwissenschaftliche Rundschau, 1951, S. 97–108.)
79. Die Kreisläufe des Wassers auf der Erde. (Schriften d. Naturwiss. Vereins f. Schleswig-Holstein, Bd. XXV, Festschr. K. Gripp, Kiel 1951, S. 185–95.)
80. Die größten Tiefen des Weltmeeres in kritischer Betrachtung. (Die Erde. Zeitschr. d. Ges. f. Erdkunde zu Berlin, Bd. II, 1950/51, H. 3/4, S. 203–14.)
81. Gesetzmäßige Wechselbeziehungen zwischen Ozean und Atmosphäre in der zonalen Verteilung von Oberflächensalzgehalt, Verdunstung und Niederschlag. (Archiv f. Meteorologie, Geophysik u. Bioklimatologie, Ser. A, Bd. 7, Wien 1954, S. 305–28.)
82. (Zusammen mit *W. Brogmus* und *E. Noodt*:) Die zonale Verteilung von Salzgehalt, Niederschlag, Verdunstung, Temperatur und Dichte an der Oberfläche der Ozeane. (Kieler Meeresforschungen, Bd. X, H. 2, 1954, S. 137–161.)
83. (Zusammen mit *W. Brogmus*:) Ozeanographische Ergebnisse einer Untersuchungsfahrt mit Forschungskutter "Südfall" durch die Ostsee Juni/Juli 1954. (Kieler Meeresforschungen, Bd. XI, H. 1, 1955, S. 3–21.)
84. Stromgeschwindigkeiten im Tiefen- und Bodenwasser des Atlantischen Ozeans... (Papers in Marine Biology and Oceanography; Bigelow Volume Deep Sea Research. London 1955, S. 373–97.)
85. Der heutige Stand der Tiefseeforschung. Eine Übersicht über die bisher erreichten Maximaltiefen. (Die Umschau, 56. Jg., H. 22, 1956, S. 673–75.)
86. Das Institut für Meereskunde der Universität Kiel. (Kieler Meeresforschungen, Bd. XII, H. 2, S. 127–53.)
87. Stromgeschwindigkeiten und Strommengen in den Tiefen des Atlantischen Ozeans unter besonderer Berücksichtigung des Tiefen- und Bodenwassers. (Wiss. Erg. Deutsch. Atl. Exped. "Meteor" 1925/27, Bd. VI, 2. Teil, 6. Lfg.; 160 S., 6 Abb., 36 Taf. u. 1 Beil., Berlin 1957.)

88. Ergebnisse eines hydrographisch-produktionsbiologischen Längsschnitts durch die Ostsee im Sommer 1956. Teil I: Die Verteilung von Temperatur, Salzgehalt und Dichte. (Kieler Meeresforschungen, Bd. XIII, H. 2, S. 163–85.)

89. Die Stromgeschwindigkeiten und Strommengen in der Atlantischen Tiefsee. (Geolog. Rundschau, Bd. 47, 1958, S. 187–95.)

90. Alexander von Humboldts Stellung in der Geschichte der Ozeanographie. (Festschrift zur A. v. Humboldt-Feier in Berlin am 18. und 19. Mai 1959, S. 90–104, Berlin 1959, S. 90–104.)

91. Remarks on the circulation of the intermediate and deep water masses in the Mediterranean Sea and the methods of their further exploration. (Annali Istituto Univ. Navale, Vol. 28, Neapel 1959; 12 S.) Auch in italienischer Sprache.

92. Sulle componenti del bilancio idrico fra atmosfera, Oceano e Mediterraneo. (Annali Istituto Univers. Navale, Vol. 28, Neapel 1959.)

93. Proposed International Indian Oceanographic Expedition 1962–1963. (Deep Sea Research 1960, Vol. VI, No. 3; "Letter to the editors", S. 245–49.)

94. Die Tiefenzirkulation des Mittelländischen Meeres in den Kernschichten des Zwischen- und des Tiefenwassers. (Deutsche Hydrogr. Zeitschr., Bd. 13, 1960, H. 3, S. 105–31, Taf. 3–9.)

95. Der Wasserhaushalt des Mittelländischen Meeres und der Ostsee in vergleichender Betrachtung. Rivista Geofisica 21, Genova 1952.

96. Über die Abnahme des Salzgehalts im Suez-Kanal von 1869–1937. Erdkunde B.V., S. 241–43, Bonn 1951.

97. Das Bodenwasser und die Vertikalzirkulation des Mittelländischen Meeres. Deutsche Hydrographische Zeitschrift, 14, 3, S. 81–92, Hamburg 1961.

98. Tables for rapid computation of potential temperature (Technical Report CU-9-61–AT (30-1) 1808 Geol. Lamont Geological Observatory.) Jan. 1961.

99. On the vertical circulation of the Mediterranean Sea. Journal Geophysical Research. Vol. 10, 163–167, London–New York 1963.

100. On the stratification and circulation of the cold water sphere of the Caribbean-Antillean basins. Deep-Sea Research, Vol. 10, 163–167, London–New York 1963.

101. Repräsentative Tiefsee-Expeditionen und Forschungsschiffe 1873–1960. Naturwissenschaftl. Rundschau, Vol. 16, 6, S. 211–214, 1963.

102. The major deep-sea expeditions and research vessels 1873–1960. A contribution to the history of oceanography. Progress in Oceanography, Vol. 2, p. 1–52, London–New York 1964.

103. Albert Defant achtzig Jahre alt. Naturwissenschaften, 51, 13. Berlin 1964, S. 301–303.

104. Albert Defant zum 80. Geburtstag. Beiträge zur Physik der Atmosphäre, 37, 2, Frankfurt a. M. 1964, S. 59–68.

105. Stratification and circulation in the Antillean-Caribbean basins. Vema-Research Series No. 2. Columbia University Press, New York 1964, p. 1–201.

106. Wasser- und Wärmehaushalt und Zirkulation in der Warmwassersphäre des Karibischen Meeres. Kieler Meeresforschungen, Bd. XXI, 1, S. 3–11, 1965.

107. Zur Frage stationärer Verhältnisse in der Makrostruktur der Kaltwassersphäre des Atlantischen Ozeans. Kieler Meeresforschungen, 1965, XXI, 1, S. 12–21.

108. Zum Gedenken an die Ausfahrt des Forschungsschiffes "Meteor" und an den Todestag von Alfred Merz vor 40 Jahren. Deutsche Hydrographische Zeitschrift 18, 1965, 4, S. 178–180.

109. Water masses and circulation in the western South Atlantic. Anals de Academia Brasileira de Ciencias. Vol. 37, Suppl. Rio de Janeiro 1965.

110. History of the investigations about the longitudinal deep-sea circulation (1800–1922). Bulletin Institut Océanographique Monaco, Numéro Special 2, 1968, S. 109–120.
111. Lectures on problems of physical oceanography. Part A (Statics), page 1–100, 1960–1961.
112. Dgl. Part B (Dynamics), page 1–67, 1962–63, erschienen in Umdrucken in "Technical Reports" of Lamont Geological Observatory (Columbia University).

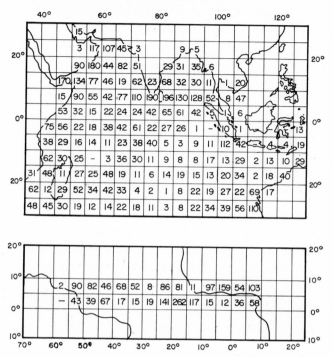

Figure 1 Upper part: Distribution of sea surface temperature observations over the Indian Ocean for June 1963. The number in each 5° square indicates the number of observations collected and processed during the International Indian Ocean Expedition (1963–1964). The distribution is typical for all months of the IIOE period. Lower part: Typical distribution of data for June 1965 for the investigated area in the Atlantic Ocean

more frequent during the first of these two years, since the Equalant 3 Expedition took place in 1965.

The available ship observations have severe shortcomings, such as: they are nonsynoptic and rather randomly distributed over a time span of one month; they are often incomplete and do not uniformly cover the area. For the purpose of time series analyses, longer series would have been desirable. In order to complement the ship data it was attempted to determine whether temperature observations by satellite could be used for a similar analysis as performed with ships' observations. It turned out that it was not feasible to use satellite data extending across the whole ocean because of a high percentage of cloud coverage. Nevertheless, these data can be used for detailed investigations of limited areas, such as the region of the Somali Current (see Section 5).

4 ANALYSIS OF THE DATA

In order to increase the reliability of the results, it was thought advisable to employ three different, independent methods of analysis.

Graphical integration

The first method consisted of a graphical integration of 24 monthly charts of the surface temperature from 1963 to 1964, which were computer contoured using an especially developed plotting routine. The contouring was based on mean values for each 1° square. Two examples of the obtained charts are presented in Figures 2 and 3. A planimeter was employed to integrate all

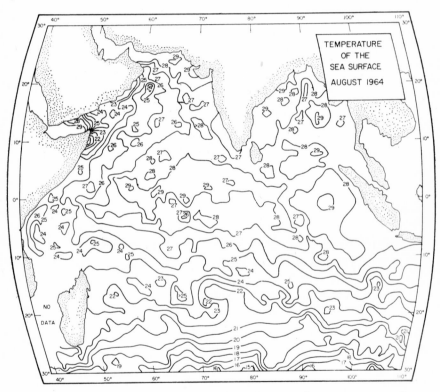

Figure 2 Computer-contoured distribution of sea surface temperatures for August 1964

closed isotherms in each monthly chart, choosing the 27°C, 28°C, and 29°C isotherms. It should be noted that this approach is only defensible for equatorial regions where weak mean temperature gradients occur. It cannot be applied in areas with large mean gradients as, for example, south of 15° S

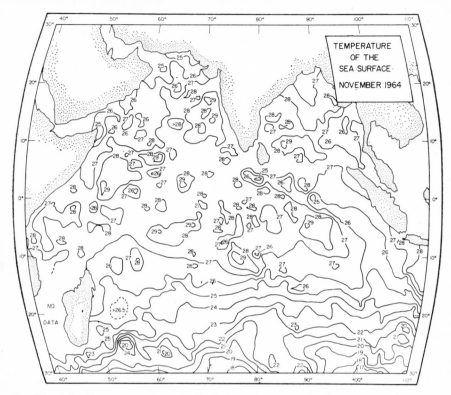

Figure 3 Computer-contoured distribution of sea surface temperatures for November 1964

in the Indian Ocean. The resulting values of the integration were sorted into conveniently chosen size classes. The smallest size class, 1, corresponds to an area of approximately 15,000 km² or a circular structure of 140 km diameter. Figure 4 shows a presentation of frequency of occurrence during all 24 months versus the various size classes. A significant peak is found in size class 10–20, corresponding to diameters varying from 440 km to 625 km. When comparing the temperature charts, the assumption of circular structures turns out to be an approximation because of the frequent occurrence of structures elongated in north–south direction and possessing elliptical shape. Additional peaks can be identified between 940 km and 1130 km and between 1300 km and 1400 km. An attempt was made to investigate the seasonal distribution of various size classes: Figure 5 presents the time-dependent distribution of all structures with a diameter smaller than 400 km and all structures with a diameter greater than 1000 km. There is evidence that the smaller structures occur preferably during the transition seasons, spring and fall, whereas the larger structures occur during winter and summer.

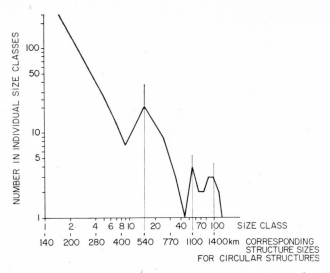

Figure 4 Distribution of size classes based on the graphical integration of all closed 28 °C isotherms occurring in monthly charts between January 1963 and December 1964

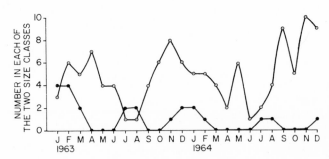

Figure 5 Seasonal distribution of two size classes: ⊙ structure sizes with diameter < 400 km, ● structure sizes with diameter > 1000 km

In addition to the graphical integration, structure functions and energy spectra were computed. In each case data from both oceans were treated in the same way. The analysis was based on the assumption that structure sizes vary more in east–west direction than in north–south direction in equatorial latitudes, hence, two narrow zonal strips were selected in each ocean extending from the equator to 4° N and from 6° N to 10° N. The results for the strip from 6° N to 10° N are presented in Table 1, but not in graphical form. Structure functions (see Wyrtki, 1967) and energy spectra were computed for temperatures averaged over four latitudinal degrees in each individual strip and each individual month.

cate considerable monthly and seasonal variation of energy. The original spectra, in particular, indicate that large peaks occur in both oceans during the winter only.

5 DISCUSSION

The results of the analysis summarized in Table 1 demonstrate that certain structures in the thermal pattern of the sea surface occur more frequently than others. Despite a relatively wide spread in the range of sizes, three different methods of analysis lead to concurring results. Sizes in both zonal strips, 0° to 4° N and 6° N to 10° N, center around mean values of 150 km, 420 km and 1200 km. The latter value being less safely established than the smaller values. It is surprising that the mean structure sizes are so similar in both oceans. This could mean that the analyzed structure sizes are atmospherically induced by weather phenomena in both areas (e.g. net cooling due to the passage of fronts) and that there exists no connection between these structures and the baroclinic oceanic circulation. This possibility can be ruled out for the structure sizes in the neighborhood of 150 km and 420 km on the grounds that they agree well with other independent observations which are known to be linked to the circulation.

The surface topography of the Arabian Sea for summer 1963 displays several eddy structures extending over 4° longitude or 440 km (Düing, 1970, Figure 14). The anticyclonic eddy near the Somali coast is well developed and seems to be a characteristic feature of the western boundary current along this coast during the summer monsoon [see also Figures 2 and 3 in Swallow *et al.* (1966) and Figure 14 in Warren *et al.* (1966)]. This westernmost eddy is clearly present in the pattern of sea surface temperature distribution observed by Nimbus 2 (Figure 9). The isotherms between 7° N and 11° N and between 51° E and 55° E roughly display a circular distribution of similar dimensions like the eddy deduced from ship observations. A second thermal structure of comparable magnitude is indicated to the northeast (southeast of the island of Socotra) of the boundary eddy. It also agrees well with the dynamic topography given by Swallow *et al.* (1966). Smaller structures in the neighborhood of 150 km occur quite frequently in the temperature distribution. Dynamic topographies do not allow a comparison in this case because of the wide spacing of hydrographic stations.

Further evidence for the structure size around 420 km comes rather unexpectedly from biological observations in the Indian Ocean: When plotting chlorophyll observations along 5° N and along 10° N through the Arabian Sea, one finds peak distances in the range of 300 km to 450 km (Figure 10). An immediate explanation for this distribution cannot be given; however,

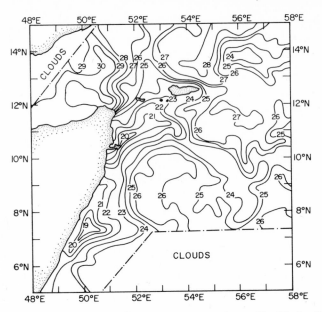

Figure 9   Distribution of sea surface temperature, as observed by Nimbus 2, July 3, 1966; contours somewhat smoothed

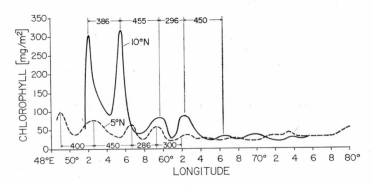

Figure 10   Distribution of vertically integrated (0–200 m) chlorophyll along two sections in the Arabian Sea. Peak distances in kilometers. Data according to *Laird, Breivogel and Yentsch* [1964]

the possible biological implications certainly merit further detailed investigation.

The similarity of the mean structure sizes in the Indian Ocean and in the Atlantic Ocean may have the following qualitative explanation: It is true that the large-scale circulation in the Atlantic Ocean is of a far more stationary behavior than the circulation in the Indian Ocean. One has to take into account, however, that approximately one-third of the Atlantic Ocean area under discussion has a monsoonal climate. (With respect to a comparison of areas possessing different conditions, the choice of the equatorial Atlantic Ocean was not very appropriate!) Like the Arabian Sea, the Gulf of Guinea shows a characteristic reversal from northeast winds during winter to southwest winds during the summer. Hence, the mechanisms of wind-driven ocean circulation can be expected to be similar in both oceanic areas. This view is supported by the fact that the maximum peaks in both chronospectra (Figures 7a and 8a) for the Atlantic Ocean and Indian Ocean occur only during winter time.

The normalized chronospectra (Figures 7b and 8b) reveal considerable fluctuation of structure sizes over time spans of typically two to three months. Fluctuations of even shorter periods seem likely but were, of course, not detectable using monthly observations. Further conclusions on the seasonal fluctuations of structure sizes cannot be drawn at the moment, since the different methods do not lead to uniform results. The normalized spectra hardly confirm the results of the graphical integration. A denser sequence of observations over longer time spans is needed to ascertain the trend indicated in Figure 5; that is, smaller sized structures to be more frequent during the transition seasons than during winter and summer. More adequate data will become available in the near future using remote sensing techniques from satellites. During the course of the present study, special efforts were made to investigate this possibility. The conclusions may be summarized as follows: A detailed discussion by Warnecke *et al.* (1969) reveals that the infrared observations by Nimbus 2 require large corrections for atmospheric water vapor and CO_2 absorption. Due to the varying and in most cases unknown amount of water vapor, an interpretation of these observations requires extreme caution, even if orbits under apparently cloud-free conditions have been selected. Such conditions rarely exist over large oceanic areas; a careful investigation of all Nimbus 2 orbits yielded only one day free of clouds over the whole Arabian Sea between the African and Indian coast in near-equatorial latitudes. For this reason a similar analysis as described above was not performed for oceanwide temperature distributions observed by satellite. For the investigation of rapid variations of small-scale phenomena, however, remote sensing techniques offer unique opportunities. The Somali coast and the oceanic area around Socotra show a high percentage of cloud-free days. Figure 9 presents an example of satellite observations; it strikingly demonstrates the superiority of this method, providing more details and higher resolution than ship observations. The oceanographer should keep in mind that the present satellite technology cannot yield reliable absolute temperatures of the sea surface, but large horizontal temperature gradients as they exist in the Somali Current and the nearby located upwelling areas can be reliably detected. Local structure sizes and their time-dependent fluctuations can be easily determined. A particularly attractive subject in terms of the dynamics of ocean currents would be a detailed investigation of the response of the Somali Current to the onset of the Southwest Monsoon and its short period fluctuations. For this purpose the seasonal fluctuations of horizontal temperature gradients would have to be determined, a task easily accomplished when using information from several years of existing and forthcoming satellite data. The value of these observations lies in the instantaneous and detailed coverage of a large area. The inherent inaccuracy of the data may not always warrant an involved mathematical analysis. An evaluation in terms of pattern recognition, such as the graphical integration of structure sizes, may be preferable for certain oceanographic applications.

6 CONCLUSIONS

The surprising similarity of structure sizes in the Indian Ocean and the Atlantic Ocean leads to the conclusion that the generating mechanisms must be similar in both monsoonally affected areas. It therefore seems feasible to apply the Rossby wave mechanism in both cases. Building on the present work, it is planned to investigate the dispersion relations for the barotropic and baroclinic case to attempt an interpretation of the various structure sizes. It is expected that the short period fluctuations of the monsoonal winds have considerable influence on the generation of the surface circulation as well as on the formation of the horizontal temperature gradients. The present results offer the possibility to explain structure sizes in terms of a spectral output function, resulting from the interaction of a conveniently defined response function with the spectral input of the monsoonal wind stress.

Acknowledgments

The author is obliged to the National Oceanographic Data Center, Washington, D.C., to the National Weather Records Center in Asheville, North Carolina and to the Goddard Space Flight Center in Beltville, Maryland for providing the data. Dr. K. H. Szekielda provided valuable assistance during the search through the Nimbus 2 observations, and D. Johnson assisted with the numerical analysis of the large amount of data. The support of this investigation, by the Office of Naval Research under Contract NONR 4008(02), is gratefully acknowledged.

References

Brumbach, R.P., Digital computer routines for power spectral analysis, Tech. Rept. TR 68-31 AC Electr.-Def. Res. Lab., S. Barbara, Calif., 1968.

Düing, W., The monsoon regime of the currents in the Indian Ocean, East-West Center Press, Univ. of Hawaii, Honolulu, 1970.

Ibbetson, A. and N.Phillips, Some laboratory experiments on Rossby waves in a rotating annulus, *Tellus*, 19 (1), 81, 1967.

Laird, J., B.B.Breivogel, and C.S.Yentsch, The distribution of chlorophyll in the western Indian Ocean during the southwest monsoon period July 30–November 12, 1963. Woods Hole Oceanographic Inst. Rept. Ref. No.64-33, 1964.

Lee, T.N., Sea surface temperature as related to circulation in the Gulf of Mexico. Unpubl. Master's Thesis, Fla. State Univ., Tallahassee, 1967.

Lighthill, M.J., Dynamic response of the Indian Ocean to the onset of the southwest monsoon. *Phil. Trans. Roy. Soc.*, A, 265, 45, 1969.

Miller, F.R. and C.Jefferies, Mean monthly sea surface temperatures of the Indian Ocean during the International Indian Ocean Expedition, Hawaii Inst. Geophys., HIG 67-14, 1967.

Miller, F.R. and R.N.Keshavamurthy, Structure of an Arabian Sea summer monsoon system, East-West Center Press, Univ. of Hawaii, Honolulu, 1968.

Nimbus Project, Nimbus 2 User's Guide, Goddard Space Flight Center, Greenbelt, Maryland.

Phillips, N., Large-scale eddy motions in the western Atlantic, *J. Geophys. Res.*, 71 (16), 3883, 1966.

Swallow, J.C. and J.G.Bruce, Current measurements off the Somali coast during the southwest monsoon of 1964, *Deep-Sea Res.*, 13, 861, 1966.

Veronis, G. and H.Stommel, The action of variable wind stresses on a stratified ocean, *J. Mar. Res.*, 15, 43, 1956.

Warnecke, G., L.M.McMillin, and L.J.Allison, Ocean current and sea surface temperature observations from meteorological satellites, Goddard Space Flight Center, Greenbelt, Maryland, NASA TND-5142, 47 p., 1969.

Warren, B., H.Stommel, and J.C.Swallow, Water masses and patterns of flow in the Somali basin during the southwest monsoon of 1964, *Deep-Sea Res.*, **13**, 825, 1966.

Warsh, K.L., E.Z.Stakhiv, and M.Garstang, On the relation between the surface temperature of the Gulf of Mexico and its circulation, 1970 (in press).

Wyrtki, K., Spectrum of ocean turbulence over distances between 40 to 1000 kilometers, *Deut. Hydro. Zeit.*, **20** (4), 176–186, 1967.

With these assumptions, or approximations, it follows that

$$\frac{\partial \bar{S}}{\partial t} = \left[\left(K \frac{\partial S}{\partial z} \right)_h - \left(K \frac{\partial S}{\partial z} \right)_0 \right] \frac{1}{h} - (w_h S_h - w_0 S_0) \frac{1}{h}$$

$$- \left(\bar{u} \frac{\partial \bar{S}}{\partial x} + \bar{v} \frac{\partial \bar{S}}{\partial y} \right) + \frac{\bar{S}}{h} (w_h - w_0) \tag{9}$$

The terms "sea surface" and "sea surface salinity" are not clearly enough defined. If we refer to a sea surface salinity, S_0, it is understood that this salinity value applies to a finite volume of water *near* the sea surface. How big or small this volume is, depends on the techniques by which "surface" water is sampled. The exact salinity at the air-sea interface may differ from the "surface salinity", S_0, as determined by present means. For example, with a completely calm sea surface and no horizontal motion, evaporation may produce an extremely thin layer of high surface salinity, and strong vertical (negative) salinity gradients. If a maximum negative density gradient is reached, vertical convection and mixing starts (Neumann, 1948). Almost the opposite is true when rain falls at the sea surface. With its impact, each rain drop produces some stirring in a thin surface layer. This helps to mix fresh water with sea water more rapidly than molecular diffusion would do without stirring.

For such reasons, the subscript 0 in the preceding equations does not apply to the true air-sea interface ($z = 0$), but rather to a level at a short distance below the sea surface. This may be 0.5 or 5 meters below the sea surface, depending on the techniques of sampling the "sea surface" salinity.

With the assumption that the vertical eddy diffusion coefficient, K, is approximately the same at the upper and lower boundaries of the layer h, the terms containing w_h and w_0, respectively, can be rearranged to obtain eq. (10).

$$\frac{\partial \bar{S}}{\partial t} = K \left[\frac{(\partial S/\partial z)_h - (\partial S/\partial z)_0}{h} \right] - \frac{w_h}{h} (S_h - \bar{S}) + \frac{w_0}{h} (S_0 - \bar{S})$$

$$- \left(\bar{u} \frac{\partial \bar{S}}{\partial x} + \bar{v} \frac{\partial \bar{S}}{\partial y} \right) \tag{10}$$

The term, $w_0 (S_0 - \bar{S})$ [gr cm^{-2} sec^{-1}] can be considered as a "source" or a "sink" for the salt amount per unit area and time at the sea surface. With the assumption that this term is proportional to $E - P$,

$$\frac{w_0}{h} (S_0 - \bar{S}) = b (E - P) \tag{11}$$

If $E - P$ is given in centimeters per unit time, b has the dimensions gr cm^{-4}. Bortkowski (1967) assumes $w_0 = P - E$ and $b = \bar{S}/h$ with the additional assumption that $(K \, \partial S/\partial z)_0 - (wS)_0 = 0$, where the subscript 0 refers exactly to the air-sea interface. The homogeneity of this equation reflects the assumption that the salt flux through the sea surface is equal to zero. In eq. (10), the subscript, 0, or $z = 0$, refers to an upper boundary of the volume which is slightly below the sea surface. With the assumptions made by Bortkowski (1967), eq. (10a) would follow

$$\frac{\partial \bar{S}}{\partial t} = \frac{K}{h} \left(\frac{\partial S}{\partial z} \right)_h - \frac{w_h}{h} (S_h - \bar{S}) - \frac{w_0}{h} \bar{S} - \left(\bar{u} \frac{\partial \bar{S}}{\partial x} + \bar{v} \frac{\partial \bar{S}}{\partial y} \right) \qquad \text{(10a)}$$

The unknown quantity w_h can be determined only if the horizontal current distribution at the level h is adequately known. Also, the last term in eq. (10), or (10a), is very difficult to evaluate because it requires an accurate and detailed knowledge of both the horizontal distribution of currents and salinity at the level of $\bar{S}$. In most oceanic regions, this information is either inadequate or lacking.

In the following, an attempt is made to apply eq. (10) to observations in the area bounded by 2° N to 2° S and 7° E to 9° E, neglecting possible horizontal advection and the term $w_h (S_h - \bar{S})/h$.

3 THE ANNUAL MARCH OF SALINITY IN THE UPPER LAYER

As a result of concentrated efforts during the International Co-operative Investigations of the Tropical Atlantic (ICITA), and subsequent oceanographic cruises into the Gulf of Guinea, knowledge of oceanographic conditions in this part of the tropical Atlantic Ocean has significantly improved. An attempt can be made to study mean monthly surface and subsurface salinity variations in different parts of this region and to relate these variations to corresponding variations of evaporation, precipitation and vertical mixing.

The salinity data used in this study is based on data collected by the National Oceanographic Data Center (NODC) and is supplemented by observations in the area not yet filed by this agency. The data are presented in Table 1 along with the number of observations used to compute monthly means for each depth. Similar data for deeper layers are also available and under investigation, but are not included in this report. In Table 2, mean monthly salinities for the combined fields, 0° to 2° N, 0° to 2° S, 7° E to 9° E are shown. These are weighted averages, according to the number of stations in the northern and southern fields, respectively. Salinities for July are inter-

Table 1 Mean salinity ($^o/_{oo}$) for 0°–2° N, 7°–9° E and number of stations

Depth (m)	J	F	M	A	M	J	J	A	S	O	N	D
0	28.51 (1)	29.72 (1)	31.10 (8)	32.33 (5)	30.68 (15)	31.88 (14)		33.38 (13)		34.48 (2)	30.78 (14)	
10	31.49 (1)	29.74 (1)	31.69 (8)	32.50 (5)	31.03 (4)	32.06 (13)		33.12 (13)		34.48 (2)	31.50 (10)	
20	33.21 (1)	34.58 (1)	33.38 (8)	34.77 (5)	33.21 (4)	33.66 (13)		33.28 (13)		34.52 (2)	32.90 (10)	
30	34.46 (1)	35.50 (1)	34.55 (8)	35.09 (5)	35.97 (5)	35.75 (13)		34.70 (13)		34.50 (2)	33.85 (9)	
50	35.58 (1)	35.81 (1)	35.92 (8)	35.98 (5)	36.09 (5)	35.91 (13)		35.80 (13)		34.76 (2)	35.80 (9)	
75	35.73 (1)	35.84 (1)	35.87 (7)	36.03 (5)	36.06 (4)	35.78 (13)		35.79 (13)		35.74 (2)	35.83 (9)	

Table 1 (cont.)

Mean salinity ($^o/_{oo}$) for 0°–2° S, 7°–9° E and number of stations

Depth (m)	J	F	M	A	M	J	J	A	S	O	N	D
0		31.79 (8)	32.00 (11)	33.22 (12)	32.14 (18)	34.24 (7)		33.80 (11)	33.97 (7)		31.81 (24)	32.36 (1)
10		32.16 (8)	32.75 (5)	33.27 (12)	32.63 (8)	35.12 (7)		34.33 (10)	34.41 (7)		34.37 (7)	33.14 (1)
20		33.27 (8)	34.22 (4)	34.00 (12)	34.30 (7)	35.89 (6)		35.37 (10)	35.04 (7)		34.69 (7)	33.96 (1)
30		35.33 (7)	34.76 (4)	35.39 (12)	35.85 (7)	36.01 (6)		35.67 (10)	35.54 (7)		34.86 (7)	35.00 (1)
50		35.88 (7)	36.24 (6)	35.99 (12)	36.08 (6)	35.99 (6)		35.82 (10)	35.92 (4)		35.64 (7)	35.75 (1)
75		35.89 (7)	35.99 (4)	36.05 (12)	35.99 (4)	35.88 (6)		35.79 (10)	35.86 (7)		35.78 (6)	35.75 (1)

Island observations of rainfall may be misleading. They can be used only after critical examination and reduction of measurements to most probable amounts in the open sea. A long list of references could be given concerning this reduction (Wüst, 1922, 1936, 1954). Reduction factors may differ from island to island, depending on the location of the rain gauge, the size, the geographical position, and the topographical conditions around the island. They depend also on meteorological conditions around the island such as the prevailing wind direction, which can change with the season. For such reasons, the reduction factor for an island may even change from season to season.

In spite of these difficulties, an attempt is made to make the best possible estimate for mean monthly amounts of rainfall in the area under consideration. This area is between the island of Sao Tomé (capital Sao Tomé at 0° 20′ N, 6° 43′ E) and the town of Libreville (0° 23′ N, 9° 26′ E) on the African coast. For both stations, reasonably long records of rain measurements are available (H.O. Pub. 105, pp. 276–277). Mean monthly amounts of rainfall in cm/month, and number of days per month with 0.025 cm or more rain are given in Table 5. The mean monthly rainfall at Libreville is 2.38 times higher than the rainfall at Sao Tomé. The annual march for both stations is shown in Figure 3, and it is seen that the trend from month to

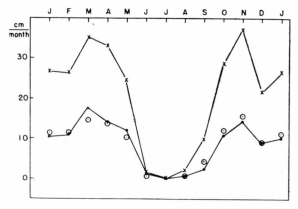

Figure 3 Mean monthly rainfall (cm/month) at Sao Tomé (small circles) and Libreville (crosses). For comparison with the rainfall at Sao Tomé, ☉ shows the rainfall at Libreville reduced by the factor 0.42.

month at both stations is similar. This is even better illustrated if the mean monthly rainfall at Libreville is divided by the factor 2.38 and plotted in Figure 3 for comparison with the rainfall at Sao Tomé. The annual march of rainfall at both stations is almost identical if the amount observed at Libreville is reduced by a factor of 0.42. In June, July, and August the rainfall

Table 5   Rainfall, P (cm/month), and number of days (N) per month with 0.025 cm or more rain. P_{ov} represents overlapping means of P for Sao Tomé

		J	F	M	A	M	J	J	A	S	O	N	D	Yearly monthly average
Sao Tomé	P	10.4	10.9	17.8	14.2	12.0	1.3	0.03	1.0	2.3	10.9	14.5	9.2	8.7
	N	10	8	11	11	8	1	–	2	5	12	10	7	
	P_{ov}	10.2	12.5	15.2	14.5	9.9	3.7	0.6	1.1	4.1	9.7	12.3	10.8	8.7
Libreville	P	26.7	26.4	35.0	33.0	24.4	1.6	0.05	2.0	9.9	28.7	37.1	21.8	20.6
	N	12	14	17	16	11	3	–	5	11	19	18	10	

at both stations is very close to zero as opposed to the two pronounced maxima in March–April and October–November.

Information about rain frequency in the inner Gulf of Guinea, based on data from ship reports, was made available by the National Geophysical Data Center, ESSA, Asheville, N.C. In Figure 4, the rain frequency (%) for

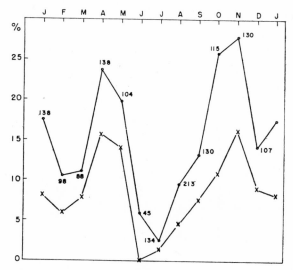

Figure 4 Frequency (%) of rain in the ocean area 2° N to 2° S, 6° E to 10° E with the total number of observations per month. All types of rain are shown by small circles, and moderate and heavy rain (omitting light rain) by crosses

each month over the ocean area 2° N to 2° S, 6° E to 10° E is shown for moderate and heavy rain and for all types of rain, together with the total number of observations (number of samples). These numbers are relatively small, and not much significance can be placed on the details of these presentations. However, the mean monthly variation of rain frequency in this area is similar to the annual march of rainfall shown in Figure 3, with the exception of March. The low value of 12 % frequency for all types of rain in March (Figure 4) is based on only 88 samples. To illustrate this correlation, Figure 5a shows the correlation between rainfall (*P* cm/month) at Sao Tomé and rain frequency (%) in 2° N to 2° S, 6° E to 10° E. For comparison, Figure 5b correlates rainfall at Sao Tomé with its own number (*N*) of days per month with a rainfall $P \geqslant 0.025$ cm. A perfect correlation between *P* and *N* cannot be expected because the average amount of rain in a month is not exactly physically related to the number of showers or rainy days in the same month. Because of this fact, and in view of the scarcity of observations over this part of the ocean, the correlation in Figure 5a can be considered fairly good and as a guide to estimate the rainfall over the area under consideration.

The rainfall over the ocean area between the island of Sao Tomé and the coastal station Libreville is definitely smaller than the rainfall obtained from an average of Sao Tomé and Libreville. For the ocean area in the vicinity of Sao Tomé, the rainfall will probably be a little less than at the island. The island is about 25 miles long in the N–S direction and 15 miles wide in the E–W direction. Since there is evidence that the total amount of rain over the ocean increases from Sao Tomé eastward as the rain frequency increases, it was estimated that the mean monthly rainfall over the island of Sao Tomé is fairly representative for the area under consideration.

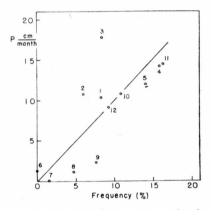

Figure 5a Rainfall at Sao Tomé (P cm/month) correlated with frequency (%) of moderate and heavy rain in the area 2° N to 2° S, 6° E to 10° E. Numbers indicate months

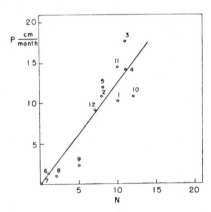

Figure 5b Rainfall at Sao Tomé (P cm/month) correlated with number of days per month (N) when the total amount of rain per day at Sao Tomé exceeded 0.025 cm

5 EVAPORATION

There are several methods for computing the evaporation from the oceans. In this paper we shall follow, essentially, the works by Sverdrup (1936, 1937–38, 1951), based on physical-hydrodynamical considerations. Although certain semi-empirical relationships have to be employed in the practical evaluation of his theory, Sverdrup's pioneering approach is most satisfactory.

For practical numerical computations, the exact formulae can be put into a simple form for moderate wind speeds

$$E = k\,(e_s - e_a)\,\bar{u} \qquad (16)$$

where E is the evaporation height per unit time, e_s the saturation water vapor pressure of air in contact with the sea surface, e_a the water vapor pressure in the air at the height a, where the mean wind speed $\bar{u}$ is measured. The factor k depends on the aerodynamical characteristics of the air flow over the sea surface which can be characterized either as a hydrodynamically "rough" or "smooth" surface, or both, in a transient state.

For wind speeds between 4 and 12 m/sec, Sverdrup *et al.* (1946) suggest a simple "working" formula

$$E\,(\text{cm/year}) = 3.7\,(e_s - e_a)\,\bar{u}$$

where e is in millibars (mbar) and $\bar{u}$ in m/sec at a height of about 6 m over the sea surface. If E is expressed in cm per month,

$$E\,(\text{cm/month}) = 0.31\,(e_s - e_a)\,\bar{u} \qquad (17)$$

From Sverdrup's further work, and his critical review of the results obtained by other investigators (Sverdrup, 1951), it seems necessary to reduce $k = 0.31$ in eq. (17) by a factor of about 0.8 if the mean wind speeds are lower than 6 to 7 m/sec. Since this is the case in the region east of Sao Tomé island (see Table 7), eq. (18) was accepted as a practical formula to compute mean monthly evaporation rates in this area.

$$E_0\,(\text{cm/month}) = 0.25\,(e_s - e_a)\,\bar{u} \qquad (18)$$

For a numerical evaluation of eq. (18), we have to know mean monthly sea surface temperatures, mean monthly wind speeds, air temperatures, and wet bulb temperatures, all measured at about the same height of 6 m over the sea surface.*

---

* Actually, one should use *simultaneous* measurements of these quantities to compute E, and then take the average of E for a month.

Table 6 A) Mean sea surface temperature (°C) in the field 2° N–2° S, 6° E–8° E, (after Böhnecke, 1938). B) Mean sea surface temperature (t_s) in the field 2° N–2° S, 7° E–9° E from NODC data as used in this report. C) Mean air temperature (t_a) in the field 2° N–2° S, 7° E–9° E from NODC data as used in this report. D) Mean wet bulb temperature in the field 2° N–2° S, 7° E–9° E corresponding to the data as shown in B) and C)

	J	F	M	A	M	J	J	A	S	O	N	D
A)	27.3	27.8	28.3	28.0	27.5	25.1	24.5	25.3	23.9	25.7	26.1	27.2
B)	28.0	28.5	28.7	28.7	28.0	25.6	(25.0)	25.5	24.2	26.6	27.2	27.0
C)	27.8	27.6	27.4	28.4	27.8	(25.6)	(24.2)	24.3	(25.7)	26.9	26.3	(26.8)
D)	(24.7)	24.8	25.4	25.8	25.6	(23.4)	(22.4)	22.6	(23.4)	24.0	24.5	24.7

Mean sea surface temperatures, recorded simultaneously with the sea surface salinities used in this study are given in row B of Table 6. For a comparison with these NODC data, row A shows the mean monthly surface temperatures in the field 2° N to 2° S, 6° E to 8° E, according to Böhnecke (1938).

Row C in Table 6 shows the mean air temperatures in the field of interest, also recorded by research vessels and taken simultaneously with sea surface temperature and salinity. Row D gives the wet bulb temperatures taken simultaneously with the air temperatures in row C.

The ships also reported wind speed. However, these observations (estimated in knots) were taken at "station time" and are not always representative for estimating the mean evaporation. For example, the observer may have made his estimate of wind speed during a gust, or while the wind just happened to die out completely at station time. Therefore, it seems better to estimate the mean monthly wind speeds from observations at Sao Tomé island at a height of about 16 feet over the sea surface. These data are shown in Table 7. The mean monthly wind speeds are low and exceed 5 m/sec only in June. The mean annual wind speed is 2.7 m/sec.

The basic data for computing E are presented in Table 7. The sea-air temperature difference, Δt, and the mean water vapor pressure difference, Δe, are computed from the data in Table 6. The value for E_0 is obtained from eq. (18).

Table 7 Evaporation (E) in cm/month estimated from the data in Table 6 and the mean monthly wind speed ($\bar{u}$ m/sec) at Sao Tomé island. $\Delta e = e_s - e_a$, $\Delta t = t_s - t_a$. The stability factor, γ, is a correction for air stability in the 6 m layer over the sea surface. $E_0 = 0.25 (e_s - e_a) \bar{u}$ (cm/month) is without correction for air stability

	J	F	M	A	M	J	J	A	S	O	N	D	Year
Δt (°C)	0.2	0.9	1.3	0.3	0.2	0.0	0.8	1.2	−1.5	−0.3	0.9	0.2	0.35
Δe (mbar)	8.1	8.9	7.8	7.3	5.8	4.9	5.2	5.7	2.5	6.3	6.0	5.3	6.15
$\bar{u}$ (m/sec)	1.8	2.2	2.4	2.6	3.3	5.1	2.9	2.9	2.4	2.2	1.8	2.2	2.65
E_0	3.6	4.9	4.7	4.7	4.9	6.4	3.8	4.1	1.4	3.5	2.7	2.9	
γ	1.2	1.3	1.4	1.1	1.0	1.0	1.25	1.3	0.1	0.7	1.5	1.1	
$E_0 \times \gamma$	4.3	6.4	6.6	5.2	4.9	6.4	4.7	5.3	0.1	2.5	4.0	3.2	
E	4.6	5.9	6.2	5.5	5.4	5.6	5.3	3.8	2.0	2.3	3.4	3.7	4.48

The effect of atmospheric stability on the evaporation rate was computed following the method outlined by Deardorff (1968), where the Monin-Obukov similarity theory is employed to express the ratio, γ, of the evaporation coefficient to its value under neutral conditions. This ratio is expressed in terms of the Monin-Obukov length which, in turn, can be related to a bulk

Table 8 Difference $E - P$ with E from Table 7 and overlapping means of P for Sao Tomζ (P_0) from Table 5 (cm/month)

	J	F	M	A	M	J	J	A	S	O	N	D	Year
E	4.6	5.9	6.2	5.5	5.4	5.6	5.3	3.8	2.0	2.3	3.4	3.7	4.5
P	10.2	12.5	15.2	14.5	9.9	3.7	0.6	1.1	4.1	9.7	12.3	10.8	8.7
$E - P$	−5.6	−6.6	−9.0	−9.0	−4.5	+1.9	+4.7	+2.7	−2.1	−7.4	−8.9	−7.1	−4.2

Richardson number (see, e.g., Cardone, 1969). The final mean monthly evaporation rates $E_0 \times \gamma$ were smoothed by overlapping means and are shown as E-values in Table 7.

The evaporation shows a minimum in September–October and a maximum in February–March. The range during the year is 4.2 cm/month, and the total evaporation per year is 54 cm. Figure 6 presents the annual march of evaporation in comparison to precipitation. During most of the year, precipitation is the dominant factor in the difference $E - P$ (see Table 8). The total yearly difference of $E - P$ is -50.4 cm/year. This value appears reasonable for this region of low wind speed and fairly high relative air humidity throughout the year (see also Jacobs, 1951). The difference of $E - P = -50.4$ cm/year must be balanced by other factors, if the annual mean of surface salinity in this region is constant.

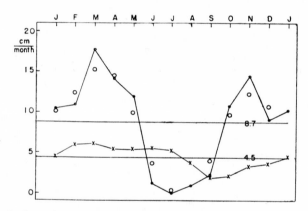

Figure 6 Annual march of rainfall at Sao Tomé recorded (small circles) and smoothed by overlapping means (large circle) in comparison with the computed annual march of evaporation (crosses) in the ocean area 2° N to 2° S, 7° E to 9° E

6 COMPUTATION OF MEAN MONTHLY VERTICAL SALINITY GRADIENTS AND THEIR DIFFERENCES WITH DEPTH

The term

$$[(\partial S/\partial z)_h - (\partial S/\partial z)_0] \frac{1}{h}$$

in eq. (10) is numerically evaluated by means of the difference form

$$\frac{S_{h+\Delta z/2} - 2S_{h-\Delta z/2} + S_{0-\Delta z/2}}{(\Delta z)^2} = \nabla(z) \qquad (19)$$

where $\Delta z = h = 10$ m. With equal distances Δz, the top of the uppermost layer of thickness h is at a water depth of 5 m ($z = 0$), the bottom at 15 m ($z = h$). $(\partial S/\partial z)_h$ is obtained from the salinities at 20 m and 10 m depth, $(\partial S/\partial z)_0$ from the values at 10 m depth and the sea surface. $\nabla(1)$ for the upper layer with an average depth at 10 m, is therefore,

$$\nabla(1) = (S_2 - 2S_1 + S_0)/10^6 \qquad (19a)$$

where $(\Delta z)^2 = 10^6$ cm^2. The salinity data for the evaluation of $\nabla(z)$ is given in Table 2. Since the surface value for January appears questionable, $\nabla(1)$ for January is interpolated.

Results of the numerical evaluation of eq. (19) are shown in Table 9a, not only for depths around 10 m, but also for 20, 30, and 40 m. In this report, use is made only of $\nabla(1)$ for 10 m. The additional values of $\nabla(z)$, however, are listed to show the continuity of the month-to-month variation of $\nabla(z)$ with depth which appears quite interesting. Overlapping monthly means of the computed values in Table 9a are shown in Table 9b. Results obtained for $\nabla(1)$ for each month are graphically presented in Figure 7. They show a surprisingly smooth annual march of $\nabla(1)$ with high values in spring and low values in fall. Slightly negative values may occur in or around November. The high values in spring are probably, to a great extent, associated with the Equatorial Undercurrent which extends during this time of the year far east-ward into the inner Gulf of Guinea (Rinkel *et al.*, 1966; Neumann, 1969). This subsurface current is characterized by high salinities at about 50 m depth along its course in the Gulf of Guinea.

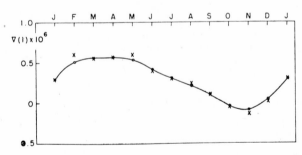

Figure 7 $\nabla(1) \times 10^6$ at 10 m depth (crosses indicate computed, small circles indi-cate overlapping means)

The vertical distribution of $\nabla(z)$ for each month (Table 9b) is shown in Figure 8. There is a significant and continuous change from month to month which indicates significant variations in the vertical salinity structure within the upper 50 meter layer. These variations are the result of combined effects of periodic intrusions of high salinity water with the Undercurrent into this

Table 9a $\nabla(z) \times 10^6$ for depths around 10 m, 20 m, 30 m, and 40 m as derived from the data in Table 2. $(\triangle z)^2 = 10^6$ cm^2

	J	F	M	A	M	J	J	A	S	O	N	D
$\nabla(1)$ 10 m	(0.60)	1.20	1.10	1.12	1.18	0.78	(0.60)	(0.48)	(0.20)	-0.08	-0.30	0.04
$\nabla(2)$ 20 m	-0.46	0.40	-0.60	-0.32	0.20	0.24	(0.32)	0.40	-0.12	-0.06	-0.30	0.22
$\nabla(3)$ 30 m	-0.70	-1.58	-0.28	-0.56	-1.82	-1.36	(-1.06)	-0.76	-0.24	-0.02	0.18	-0.60
$\nabla(4)$ 40 m	0.04	-0.18	-0.06	-0.20	-0.16	-0.40	(-0.06)	0.30	-0.14	0.14	-0.24	-0.34

Table 9b Overlapping means for $\nabla(z) \times 10^6$

	J	F	M	A	M	J	J	A	S	O	N	D	Year
$\nabla(1)$	0.62	1.02	1.12	1.14	1.06	0.84	0.62	0.44	0.20	-0.06	-0.14	0.10	0.58
$\nabla(2)$	-0.08	-0.06	-0.28	-0.26	0.08	0.24	0.32	0.24	0.02	-0.14	-0.12	-0.08	
$\nabla(3)$	-0.90	-1.04	-0.68	-0.80	-1.40	-1.40	-1.06	-0.70	-0.32	-0.02	-0.06	-0.48	
$\nabla(4)$	-0.10	-0.10	-0.12	-0.16	-0.22	-0.26	-0.06	+0.02	-0.32	-0.02	-0.16	-0.22	

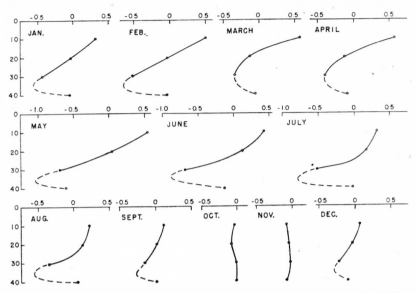

Figure 8 $\nabla(z)$ for 10, 20, 30, and 40 m depth according to eq. (19) and Table 9b
for each month of the year. The abscissa shows $\nabla(z)$ in 10^6 [gr/cm³/cm²]

region, and the effects of $E - P$ at the sea surface. It is seen that in the fall,
$\nabla(z)$ is close to zero. The effect of salinity flux by vertical mixing on salinity
changes should, therefore, reach a minimum during this part of the year.

7 COMPUTATION OF SEASONAL SALINITY VARIATIONS IN THE NEAR-SURFACE LAYER

With horizontal advection and the term containing w_h neglected, it follows
from eq. (10) and eq. (11) that

$$\frac{\partial S}{\partial t} = K \nabla (1) + b (E - P) \qquad (20)$$

$\nabla(1)$ is given by eq. (19a), where $\Delta z = 10$ m and $h - \Delta z/2 = 10$ m.

The coefficient b is proportional to the difference $S_0 - \bar{S}$. In general, b is
not constant but varies with time, if $S_0 - \bar{S}$ varies with time. Application
of eq. (20) to the near-surface layer with an average depth of 10 m requires
the evaluation of S_0 at a depth of 5 m and S_h at a depth of 15 m. $\bar{S}$ is the
average salinity in the layer of thickness, h, between 5 m and 15 m depth,
and approximately, $\bar{S} = \frac{1}{2} (S_0 + S_h) = S_1$. If the observed sea surface
salinity is used for S_0, the annual mean for the difference $S_1 - S_0 = 0.36$.
Since the annual march of S_0 is similar to the annual march of S_1, as shown

in Figures 1 and 2, it seems justified to assume that $b \propto S_0 - S_1$ is approximately constant. The correlation between S_1 and S_0 is shown in Figure 9 for each month of the year (overlapping means from Table 3). The line represents $S_1 = 0.36 + S_0$. With the exception of November, the standard deviation of the observed differences $S_1 - S_0$ from the annual mean is $0.15^0/_{00}$. Although this seems large, the deviations from the annual mean (0.36) are small when compared to the total range of about $2.4^0/_{00}$ for S_0 or S_1 during the year. The deviations from the mean value $S_1 - S_0 = 0.36$ are definitely within the limits of accuracy with which the annual march of salinity has been determined.

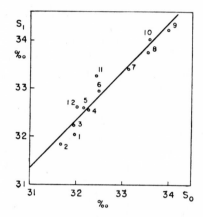

Figure 9 Mean monthly observed surface salinities, S_0, versus mean monthly observed salinities at 10 m depth, S_1. The line represents $S_1 = S_0 + 0.36$ ($^0/_{00}$). Numbers indicate months

Table 10 compiles the basic data for the computation of $\partial S_1/\partial t$ by means of eq. (20). $(\partial S_1/\partial t)_{obs}$ is from Table 4, $\nabla(1) \times 10^6$ from Table 9b, and $E - P$ from Table 8.

For the annual mean, $\partial S_1/\partial t = 0$, and with the mean $(E - P) = -4.24$ and the mean $\nabla(1) \times 10^6 = 0.58$ for the year, it follows from eq. (20) that

$$K = 7.31 \times 10^6 b$$

In order to determine b, another equation is needed. For this purpose it is assumed that at least during one month of the year the two terms on the right hand side of eq. (20) give a sufficiently accurate estimate of $\partial S_1/\partial t$ This month should be chosen where $\partial S_1/\partial t$ is not too close to zero, in order to avoid small differences on the right hand side of eq. (20), and where $(\partial S_1/\partial t)_{obs}$ can be estimated to be most accurate. A suitable month for such a "reference point" is July, where $(\partial S_1/\partial t)_{obs} = 0.96$. The balance $0.96 = 7.31 \times 0.62b + 4.7b$

yields
$$b = 0.104$$
and
$$K = 0.76 \times 10^6 \quad [\text{cm}^2/\text{month}]$$

or $K = 0.3$ [cm²/sec]. This compares well with other estimates of vertical eddy diffusion coefficients in strongly stable water stratification. For example, Montgomery (1939) obtained as a mean value for the Equatorial Counter-current in the Atlantic Ocean $K/u = 0.012$ cm, where u is the mean current speed. According to Montgomery, $u = 30$ cm/sec, and $K = 0.36$ cm²/sec in the "Sprungschicht" of this current.

With these values of K and b, the annual march of S_1 should be given by

$$\frac{\partial S_1}{\partial t} = 0.76 \, \nabla(1) \, 10^6 + 0.1 \, (E - P) \tag{21}$$

Evaluation of eq. (21) using the data from Table 10 yields $\partial S_1/\partial t$ as shown in the bottom row of this table. Harmonic analysis of these computed values leads to the result

$$\frac{\partial S_1}{\partial t} = 0.692 \sin \left(\frac{2\pi}{T} t + 305.3° \right) + 0.446 \sin \left(\frac{4\pi}{T} t + 69.7° \right) \tag{22}$$

The computed annual march of salinity, S_1, is obtained from

$$S_1 = 32.96 - 0.69 \cos \left(\frac{2\pi}{T} t + 305.3° \right) - 0.22 \cos \left(\frac{4\pi}{T} t + 69.7° \right) \tag{23}$$

The integration constant from eq. (22) is the annual mean, 32.96°/₀₀ for S_1.

Observed and computed monthly means are shown in Figure 10 and 11, respectively. Any inaccuracies in the basic data such as those in the S_1 values as determined from the observed annual march appear, of course, strongly pronounced in Figure 10, because the time derivative of S_1 multiplies the amplitude of the second harmonic of S_1 by a factor of two. Therefore, any inaccuracy of the approximation to the real annual march is accordingly magnified. Computed and observed salinities, S_1, in Figure 11 agree fairly well.

It appears that the major features of near-surface salinity variations in this area can be explained satisfactorily in terms of the effects of $E - P$ and vertical mixing. The neglected effects of advection and lateral mixing may account for some of the differences between the observed and the computed values. However, the author is more inclined to assume that the inaccuracies of the basic data are larger than the effects of the neglected terms.

Table 10 Computation of $\partial S_1/\partial t$

	J	F	M	A	M	J	J	A	S	O	N	D	Year
$(\partial S_1/\partial t)_{obs}$	-0.54	0.14	0.37	0.24	0.20	0.53	0.96	1.00	0.33	-0.66	-1.34	-1.23	0.00
$\nabla(1) \times 10^6$	0.62	1.02	1.12	1.14	1.06	0.84	0.62	0.44	0.20	-0.06	-0.14	0.10	0.58
$E - P$	-5.6	-6.6	-9.0	-9.0	-4.5	1.9	4.7	2.7	-2.1	-7.4	-8.9	-7.1	-4.24
$\partial S_1/\partial t$	-0.10	+0.10	-0.06	-0.05	0.34	0.82	0.94	0.60	-0.06	-0.80	-1.01	-0.64	0.00

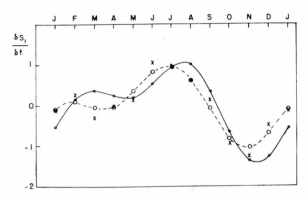

Figure 10 Observed, indicated by small circles, and computed, indicated by large circles, mean monthly values of $\partial S_1/\partial t$ in gr/cm^3 per month (or approximately $\partial S_1/\partial t\,^\circ\!/_{00}$ per month). The crosses represent the results of computation if P is taken from Table 5 without overlapping means

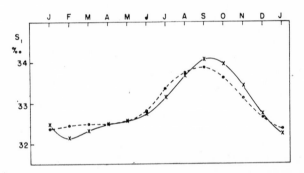

Figure 11 Observed, indicated by crosses, and computed, indicated by small circles, annual march of salinity, S_1, at 10 m depth

Acknowledgments

Thanks are extended to Dr. W. J. Pierson, Jr. for reading the manuscript of this paper, and to him and other colleagues at the Department for helpful discussions of the general problem of near surface salinity fluxes. This work was supported by the Office of Naval Research under Contract NONR 285 (57). Reproduction in whole or in part is permitted for any purpose of the United States Government.

References

Böhnecke, G., Temperatur, Salzgehalt und Dichte an der Oberfläche des Atlantischen Ozeans, *Wiss. Erg. d. Deutschen Atl. Exped. "Meteor"* **1925-27** *(Meteor-Werk)*, **5**, 2. Liefg., Die Temperatur, Berlin 1938.

Bortkowski, R. S., Methods of computing heat and salt flow in the ocean as applied to selected regions of the Atlantic Ocean (Transl. from Russian), *Akad. Nauk SSSR, Trudy Inst. Okeanol.*, **56**, *Oceanographic Research in the Atlantic*, Edit. M. V. Klenova, 1962, IPST, Jerusalem 1967.

Cardone, V. J., Specification of the Wind Distribution in the Marine Boundary Layer for Wave Forecasting, New York University, School of Engineering and Science, Geophysical Sciences Laboratory TR 69-1, Dept. of Meteorology and Oceanography, Research Division, December 1969.

Deardorff, J. W., Dependence of air-sea transfer coefficients on bulk stability. *J. Geophys. Res.*, **73**, 2549, 1968.

Defant, A., Physik des Meeres, *Handbuch der Experimentalphysik, Wien-Harms*, **25**, 2, Leipzig 1931.

Defant, A., *Physical Oceanography*, Vol. I, 164, Pergamon Press, New York, Oxford, London, Paris 1961.

Defant, A. and H. Ertel, Durch Niederschlag verursachte Störungen des Salzgehaltes im Ozean und deren Ausgleich durch Turbulenz, *Abhandl. Preuss. Akad. Wissensch.*, 10, Berlin 1939.

Jacobs, W. C., The energy exchange between sea and atmosphere and some of its consequences, *Bull. Scripps Inst. Oceanography*, Vol. 6, 27, La Jolla, Calif., 1951.

Montgomery, R. B., Ein Versuch, den vertikalen und seitlichen Austausch in der Tiefe der Sprungschicht im äquatorialen Atlantischen Ozean zu bestimmen, *Ann. d. Hydr. u. Marit. Meteorol.*, **67**, 242, Berlin 1939.

Neumann, G., Die ozeanographischen Verhältnisse an der Meeresoberfläche im Golfstromsektor nördlich und nordwestlich der Azoren. Aus den wiss. Ergebn. der Intern. Golfstromunternehmung, 1938. *Ann. d. Hydr. u. Marit. Meteorol.*, **68**, June-Beiheft, Berlin 1940.

Neumann, G., Bemerkungen zur Zellularkonvektion im Meer und in der Atmosphäre und die Beurteilung des statischen Gleichgewichts, *Ann. d. Meteorol.*, July–Aug., p. 235, Hamburg 1948.

Neumann, G., The Equatorial Undercurrent in the Atlantic Ocean, *Proc. of the Symposium on Oceanography and Fisheries Resources of the Tropical Atlantic*, pp. 33–44, UNESCO, Paris 1969.

Rinkel, M. O., P. Sund, and G. Neumann, The location of the termination area of the Equatorial Undercurrent in the Gulf of Guinea based on observations during Equalant III, *J. Geophys. Res.*, **71**, 16, 1966.

Sverdrup, H. U., Das maritime Verdunstungsproblem, *Ann. d. Hydr. u. Marit. Meteorol.*, **64**, Berlin 1936.

Sverdrup, H. U., On the evaporation from the oceans, *J. Mar. Res.*, **1** (1), New Haven, Conn., 1937–38.

Sverdrup, H. U., M. W. Johnson, and R. H. Fleming, *The Oceans*, Prentice-Hall, Inc., Englewood Cliffs, N. J., 1946 (2nd. edition).

Sverdrup, H. U., Evaporation from the oceans. *Compend. Meteorol.*, Amer. Meteorol. Soc., 1071–1081, Boston, Mass., 1951.

Wüst, G., Verdunstung und Niederschlag auf der Erde, *Zeitschr. Ges. für Erdkunde*, 1–35, 1922.

Wüst, G., Oberflächensalzgehalt, Verdunstung und Niederschlag auf dem Weltmeere, *Länderkundl. Forschung, Festschrift N. Krebs*, 347–359, Stuttgart 1936.

Wüst, G., Gesetzmäßige Wechselbeziehungen zwischen Ozean und Atmosphäre in der zonalen Verteilung von Oberflächensalzgehalt, Verdunstung und Niederschlag, *Arch. Meteorol. Geophys. Bioklim.*, *A*, 7 (Defant Festschr.), 305–328, Wien 1954.

Paper 3

Visible Oceanic Saline Fronts*

ANTHONY F. AMOS AND MARCUS G. LANGSETH, JR.

Lamont-Doherty Geological Observatory, Columbia University
Palisades, New York 10964

RUDI G. MARKL

University of Connecticut, Avery Point, Conn.

Abstract A change in sea-surface salinity of $1.1^o/_{oo}$ over a horizontal distance of less than 100 m was found some 600 km southwest of Sumatra in June 1965. This saline front was visible at the surface as a narrow band of rough water in an otherwise calm sea. A considerable amount of land-based debris was floating in the zone of turbulence, and an abundance of faunal life was noted at the front. Hydrographic studies were made across the front, using a continuously-recording salinity, temperature, depth (STD) probe. Its horizontal extent was at least 250 km, running in a NE–SW direction.

A comparison is made between this saline front and visible thermal fronts found in many parts of the world's oceans and to similar surface phenomena frequently observed in the Bay of Bengal—Andaman Sea region caused by internal waves. The relationship of the monsoon circulation of the Indian Ocean to the formation of this front is discussed. It is concluded that in June 1965 a visible boundary was formed where the eastward-flowing Equatorial Countercurrent met fresh water originating in the Bay of Bengal and Andaman Sea, due to the discharge of the great river systems emptying there.

Another possible saline front, found 800 km north of Brazil in August 1970, is discussed in light of the surface circulation in the equatorial North Atlantic and the fresh water discharge from the Amazon River.

INTRODUCTION

Japanese fishermen have recognized for many years that visible oceanic fronts are areas where fish and whales congregate (Uda, 1938). Hydrographic studies on some fronts in the open ocean have revealed that they coincide with sharp surface temperature discontinuities (Cromwell and Reid, 1956;

---

* Lamont-Doherty Geological Observatory Contribution 1622.

4 Gordon I

49

Knauss, 1957; Voorhis and Hersey, 1964; Katz, 1969; Voorhis, 1969). Voorhis and Hersey (1964) used the term "thermal front" to describe these phenomena. Most observers have found that small changes in surface salinity also occur across these fronts (Voorhis, 1969; Katz, 1969). Although water sampling programs have not been carried out on all those observed to date, Zaneveld *et al.* (1969) found that large changes in both temperature and salinity occurred at the interface of a front near the Galapagos Islands.

A strikingly visible front was found in the Indian Ocean during Cruise 9 of the *R/V Conrad*. This front was not characterized by any significant changes in thermal structure, but coincided with an abrupt change in surface salinity. More recently, another visible front was crossed during *Conrad* Cruise 13 off the north coast of South America. While no salinity measurements were made across this feature, a temperature section down to 400 m showed no structures identifiable with a thermal front. Thermal fronts occur when two water masses of different temperatures converge at the sea surface, the word" front" being borrowed from the meteorologist's lexicon. Density differences occur due to the change in temperature across the frontal boundary. However, for the front found in the Indian Ocean, the density difference is controlled entirely by salinity; hence we have called this phenomenon a "saline front". Kiyomitsu (1970) uses the term "salinity front" to describe the boundary between the shelf water and deep water in the Bering Sea. The North Atlantic Ocean front is included in this discussion because it is, by inference, a saline front.

The visible manifestation of both fronts was a narrow band of turbulent water extending from horizon to horizon. In both cases, the normal sea surface was calm in contrast to the turbulent zone, which had breaking waves with whitecaps. Both fronts were located hundreds of kilometers from land in the open ocean and deep water. The Indian Ocean front was investigated with a continuously-recording salinity, temperature, depth (STD) probe (Amos, 1966a).

The Indian Ocean front

On June 4, 1965 at 0° 41.2′ S, 93° 26.5′ E (Figure 1), a band of dark water was noticed stretching across the horizon from northwest to southeast (Figure 2). The ship's course was changed to intercept the line, and closer examination revealed a 50 to 100-m wide zone of turbulent water meandering as far as the eye could see in both directions. The turbulent zone consisted of small wavelets, 0.1 to 0.3-m high, breaking into whitecaps and apparently produced by a confluence of opposing surface currents. This area of turbulence produced a hissing sound that could be clearly heard above the ship's engine noises. Considerable biological activity was observed in this zone, mostly to the seaward side of the front. Most notable was the presence of many small (< 1-m long) sharks and an abundance of flying fish in all stages of development. In addition, jellyfish, dolphin (dorado), and a few sea snakes were observed. From a distance, sea birds were seen feeding at the front. Planktonic life was extremely rich in the zone of turbulence. A large quantity of debris, presumably land-based in origin, was floating in the frontal zone, consisting of coconut shells, small pieces of planking, sticks, leaves, a wine

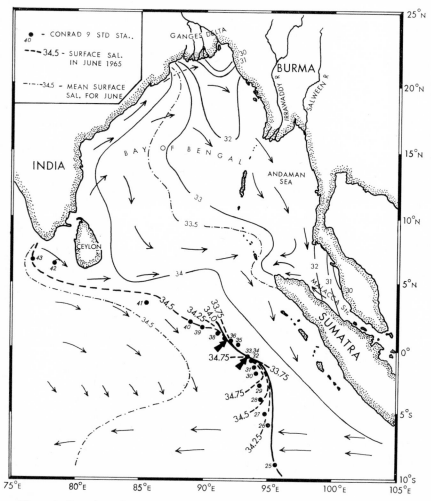

Figure 1 Location of Indian Ocean front (between solid, bold arrows). Surface salinity in June 1965 is shown by heavy contour lines. Numbered dots are *Conrad 9* STD station locations. Average June surface salinity (light contours) and surface currents (arrows) are also shown. Sources: LaViolette, 1967; H.O. Spec. Publ. 53, 1960; H.O. Publ. 566, 1970. *Conrad* stations from Amos 1966b.

bottle, Coca-Cola bottle, and a glass fisherman's float. This description is virtually identical to that of a thermal front found by Beebe (1926) off the Galapagos Islands. Meteorological conditions prevailing at the time of observation were: wind from 225° at 3–6 kts, sea calm with a 0.3-m swell from 225°, dry bulb temperature 30.0°C, wet bulb 27.8°C, barometric pressure 1012 mB, and sky generally clear.

a)

b)

Figure 2   (a) Indian Ocean front photographed from a distance of 200 m. (b) Edge of Indian Ocean front looking along axis (still photographs taken from 8 mm color movie sequence)

Profiles of salinity and temperature down to 500 m were made on either side of the front, using the STD recording continuously as the instrument was lowered and raised. The family of salinity, temperature and density curves (Figure 3) shows that to the landward side a thin (30 m) layer of

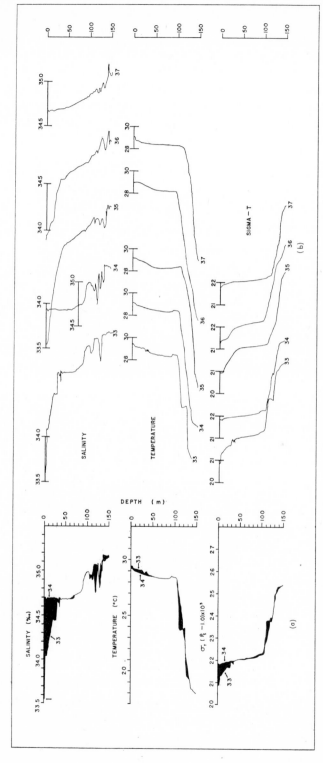

Figure 3 (a) Envelopes formed by salinity, temperature and σ_t profiles for *Conrad 9* STD Sta. 33 and Sta. 34, taken on low and high salinity sides of front, respectively. (b) Temperature, salinity and σ_t profiles from STD stations in vicinity of Indian Ocean front

relatively fresh water occurs at the surface. The surface salinity abruptly changes by 1.1°/₀₀ at the front itself. The drift behavior of the ship while these measurements were being made indicates that the front itself was moving seaward at a fairly rapid rate. The presence of land-based debris of a type that would eventually become waterlogged also suggests that the water mass had moved rapidly from land. Unfortunately, none of the man-made debris was recovered, so that its point of origin can only be surmised.

Station 33 was located approximately 100 m from the edge of the turbulent zone on the low salinity side (see Figure 4). A few minutes after the STD was lowered, the ship had drifted into the turbulent zone. Once there, however, the ship stayed in the zone for the rest of the station. The STD sea cable went

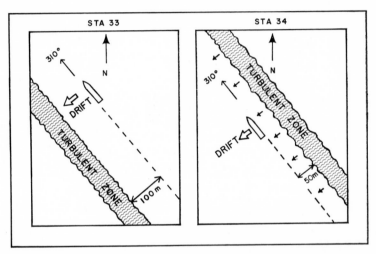

Figure 4   Position and drift of *R/V Conrad* relative to front during STD stations 33 and 34

under the ship after 35 m of wire had been payed out. As shown in Figure 4, Station 34 was located about 50 m to the seaward side of the front. During this station the front advanced on the ship. Again, the vessel remained in the front for the rest of the station. From these observations, the front seemed to be moving in the direction 210°. A rough estimate of the front's speed relative to the ship was made based on the time it took the ship to drift back into the turbulent zone. The speeds at the two stations were 5 and 15 cm/sec, respectively. However, because of the uncertainty of these observations, we feel that the speed of the front may have been grossly under-estimated. Two more stations (35 and 36) were taken on the low salinity side, neither in sight of the front. Strong surface currents were present at Station 35. A similar front was observed at 0636Z on June 5 at 0° 40′ N, 91° 40′ E, and here it was determined to be striking 330°. If the front was continuous between the two observations, as the salinity structure of Stations 35 and 36 suggest, then its length was at least 250 km. Station 37 was taken to the seaward side of this second front (Figure 3).

The North Atlantic front

This feature was found at 1131Z on 10 August 1970 at 6° 45.7′ N, 44° 46.9′ W, some 800 km from the South American continent (Figure 5). The turbulent zone here was wider (approximately 250 m) and better defined than the Indian Ocean zone. It was running east-west and stretched from horizon to horizon. Here, the turbulent zone (Figure 6) was divided into distinct zones (shown schematically in Figure 7): swells breaking in opposite direction are observed on either side of a narrow zone where the water surface appeared to be in constant agitation, giving a "frosted" appearance. Small eddies ($\sim$50 m in diameter) were noted on the edges of the turbulent zone. Although no stations or water sampling program were possible, a line of expendable bathythermographs (XBT's) was taken across the zone. The ship's behavior while crossing the zone was noteworthy. The XBT run was made at 9.5 kts from south to north. While passing through the central "frosted" zone, a distinct bump was felt on the ship, and the pit log reading jumped from 9.5 to 11.5 kts. The ship's engines were slowed because of excessive vibration. The ship's master estimated that the whole frontal zone was moving northward at about 2 kts. About 5 miles to the south, a second front was crossed, quite similar to the first in appearance, except wider (350 m) and with waves breaking to the north only. The frosted zone was still evident, and the same momentary jump in speed on crossing it (this time from 7 to 11 kts) was experienced. As with the Indian Ocean front, a hissing noise was heard in the turbulent zone. No debris was observed, nor was any unusual biological activity noted.

DISCUSSION

The Indian Ocean front

The surface circulation in the northern Indian Ocean is controlled by the monsoons, hence its seasonable variation is considerable. Surface salinity in the northeastern Indian Ocean is particularly variable due to the vast amount of fresh water discharged by the Brahmaputra–Ganges and Irrawaddy–Salween river systems during the summer monsoon and the seasonal reversal of the major surface currents. During the winter (NE) monsoon, the equatorial circulation is characterized by the North Equatorial Current flowing from east to west while the Equatorial Countercurrent brings high salinity water

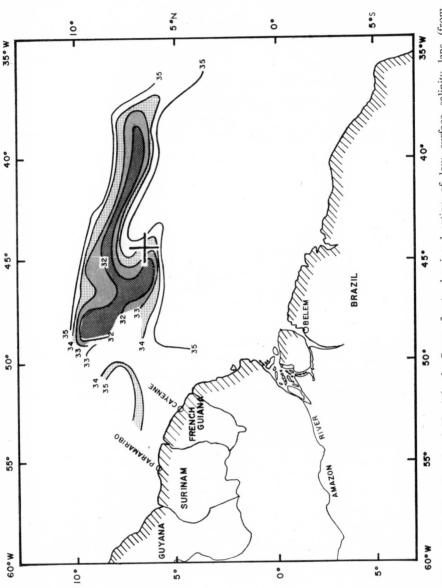

Figure 5 Position (cross) of N. Atlantic Ocean front, showing location of low surface salinity lens (from Cochrane, 1959)

a)

b)

Figure 6   N. Atlantic front photographed from: (a) Edge of turbulent zone looking southward (view is approximately normal to axis of front). Normal sea state appears just below horizon at upper center. (b) Center of turbulent zone looking along axis of front

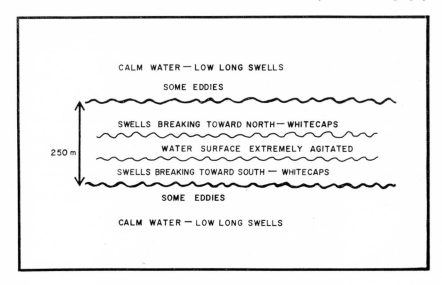

Figure 7   Schematic representation of N. Atlantic front

from the west almost to the coast of Sumatra in latitude 2°–3° S by the end of February. When the summer (SW) monsoon is fully developed, the North Equatorial Current is replaced by the SW Monsoon Current, flowing in the opposite direction, and the countercurrent apparently disappears. The South Equatorial Current, setting from east to west, is not influenced significantly.

The surface salinity distribution in early June 1965, as indicated by *R/V Conrad* STD stations, is seen to vary considerably from the mean June condition (Figure 1). The eastward displacement of the $34.5^o/_{oo}$ isopleth and the narrow region of high salinity ($>34.75^o/_{oo}$) water at about 2° S, indicate that water carried by the Equatorial Countercurrent was still present at the beginning of June 1965. At the same time, fresher water was found considerably farther south and west than is usual at this season. The May–June transition period is known to be one of high variability as the advance of the SW monsoon varies geographically from year to year (La Violette, 1967).

The dynamic topography of the sea surface compiled from data collected by ships of the International Indian Ocean Expedition (Düing, 1970) shows the Equatorial Countercurrent to have broadened into the general eastward drift in early summer. During this period, however, it is not possible to tell whether the North Equatorial Current or the Southwest Monsoon Current is present. It would appear that during this transition period fresh water from the Bay of Bengal is driven by the general clockwise circulation in the bay into the Andaman Sea and exits through the Great Passage north of Sumatra. Additional fresh water may be added by the coastal Malacca Straits current that flows through this passage (Fairbridge and Rudolfo, 1966). The absence of a well-established SW Monsoon Current may allow this fresher water to flow for a considerable distance into the Indian Ocean. A tongue of fresh water as far west as 87° E is shown in June–August (U.S. Navy Hydrographic Office, Spec. Pub. 53, 1960, Figure 148). The salinity front observed by *R/V Conrad* on June 4 and 5, 1965, may be the boundary between fresh water from the Bay of Bengal–Andaman Sea and high salinity water driven eastward by a still well-defined Equatorial Countercurrent. Whether this represents an anomalous condition or a rapidly changing seasonable occurrence is not known.

Rain gauge data from areas contributing to the river systems' discharge reported less than average rainfall for 1965 up to June, but higher than average during 1964. By early June the monsoonal rains have hardly started to contribute to the dilution of Bay of Bengal–Andaman Sea waters, thus the degree of dilution in June may reflect the previous year's precipitation.

Anomalous linear zones of agitated or rough water in the Indian Ocean have also been observed associated with internal waves. LaFond and LaFond (1968) report frequent occurrences of alternate bands of slick and rough water in the Bay of Bengal and throughout the Andaman Sea, caused by trains of internal waves. Perry and Schimke (1965) found bands of rough water in an otherwise calm sea in the Great Passage on June 13, 1964 to be caused by large-amplitude internal waves. Perry and Schimke dismiss the possibility that these surface phenomena are associated with oceanic fronts. However, the *Conrad* feature, a single band of turbulent water stretching for at least 250 km and coinciding with a large surface salinity discontinuity, appears to be an oceanic saline front rather than an internal wave related phenomenon. On the other hand, oceanic fronts may be a strong source of internal waves. Data of Voorhis and Hersey (1964) across a thermal front indicate internal waves at the thermocline.

The equatorial Atlantic front

The surface circulation of the part of the Atlantic where the front was encountered is that of the typical equatorial region without the extreme annual variation found in the Indian Ocean. The strong Guinea Current flows to the northwest along the coast of Brazil. The location of the Atlantic front (6° 46′ N, 44° 47′ W) is close to the boundary between the Guinea Current and the Equatorial Countercurrent. The countercurrent flows generally eastward in a zone about 5° wide and is found at least as far west as 50° W (Cochrane, 1969). Thus, the front is in an area of apparently diverging surface currents. The situation is complicated by the large discharge of Amazon River water into the Atlantic (Gibbs, 1970). A large tongue of fresh water exists at the surface in the Atlantic Ocean during summer, formed by Amazon River water being transported northwestward along the coast by the Guinea Current and then eastward by the countercurrent (Cochrane, 1969). This tongue is shown in Figure 5 in relationship to the location of the *Conrad* turbulent zone.

The closely-spaced series of vertical temperature profiles across this vigorous front showed only a slight depression of isotherms at the thermocline below the front. Isotherms above 50 m are essentially level. By inference we believe this is a saline front similar to that found in the Indian Ocean. In this case, the low salinity water is provided by the Amazon River discharge. Figure 5 shows the front was sighted near the edge of the tongue of low salinity water delineated by Cochrane (1969). The abrupt changes of speed of the ship while crossing the front indicate that the front was advancing at close to 2 kts or more. However, the absence of land-derived debris, so abundant in the Indian Ocean front, suggests that the low salinity Amazon water had been in the open sea for a much longer period of time than that in the Indian Ocean.

Rapidly changing sea-surface salinities in the area ($> 1^o/_{oo}$ in three hours) were detected by ships of the Barbados Oceanographic and Meteorological Expedition (BOMEX) during time-series STD measurements on anchored station (Victor Delnore, BOMAP Office, personal communication, 1970). Corresponding temperature fluctuations were of the order of only 0.1 to 0.2 °C Several such changes occurred during two weeks on station, indicating that saline fronts were passing through the area during July, 1969. Whether any of these were visible is not known.

In September 1968, a series of three visible turbulent bands was found at 8° 15′ N, 42° 5′ W by *USNS Kane* (Edward Escowitz, NavOceano, personal communication, 1970). An XBT section across these zones reveals the presence of several internal waves in the thermocline, and these are probably responsible for the surface disturbances. No similar waves were detected across the *Conrad* feature, although the spacing of the XBT observations (~ 1 km) might mask some smaller fluctuations.

SUMMARY

Visible oceanic fronts, whether thermal, saline, or both, appear to be quite common phenomena in the world's oceans, particularly in the equatorial zone. Basically, they occur when two surface water masses of different densities meet, and their boundary is visible as a narrow band of turbulent water. No matter what the parameter is that controls the density difference, descriptions of their physical appearance are remarkably similar. Their horizontal extent appears to be hundreds of kilometers, and they have maintained their identity for up to three months (Voorhis, 1969). In addition to abrupt density changes, differences in optical properties (Zaneveld *et al.*, 1969), fish fauna (Backus *et al.*, 1970), and plankton and bird life (several observers) have been found across these fronts. It would seem that they may be valuable as fishing grounds, and further investigation of nutrient levels and planktonic life associated with the fronts would seem desirable. If they prove to be of economic importance, their seasonal variation, predictability, and detection would be most useful information. Aerial spotting (Voorhis, 1969) and even detection on ship's radar (Zaneveld *et al.*, 1969) have already been accomplished. Detection by instrumented earth-orbiting satellite may also be possible and may resolve the large-scale extent and duration of these features.

Acknowledgments

We are grateful to the master of *R/V Conrad*, Capt. A.L.Jorgenson, for his most useful observations of the North Atlantic front. The *Conrad*-9 Indian Ocean work was carried out under National Science Foundation Grants G-22260 and GP-5538, and the *Conrad*-13 North Atlantic work was supported by Atomic Energy Commission Contract AT(30-1)2663, Office of Naval Research Contract N0014-67-A-010 B-004 and National Science Foundation Grant GA-1615.

References

Amos, A.F., Measurements in the Indian Ocean using a continuously recording *in situ* salinity, temperature and depth sensor (Abst.), *Trans. AGU*, **47** (1), 112, 1966a.

Amos, A.F., Physical oceanographic observations in the Indian Ocean using a continuously recording *in situ* salinity/temperature/depth sensor. *R/V Robert D.Conrad* Cruise 9, 1965, *Lamont Geol. Observ. Tech. Rep. IIOE*, 188 pp., 1966b (unpublished).

Backus, R.H., J.E.Craddock, R.L.Haedrich, and D.L.Shores, Mesopelagic fishes and thermal fronts in the western Sargasso Sea, *Marine Biol.*, **3** (2), 87–106, 1969.

Beebe, W., *The Arcturus Adventure*, G.P.Putnam's Sons, New York and London, Ch.2, pp. 41–70, 1926.

Cochrane, J.D., Low sea-surface salinity off north-eastern South America in summer 1964, *J. Mar. Res.*, **27** (3), 327–334, 1969.

Cromwell, T. and J.L.Reid, Jr., A study of oceanic fronts, *Tellus*, **8** (1), 94–101, 1956.

Düing, W., The monsoon regime of the currents in the Indian Ocean, *Int. Indian Ocean Expedition Oceanog. Monographs 1*, East-West Center Press, Honolulu, 68 pp., 1970.

Fairbridge, R.W. and K.S.Rudolfo, "Andaman Sea", pp. 32–35 in: *The Encyclopedia of Oceanography 1*, R.W.Fairbridge, ed., Reinhold Publ. Co., N.Y. 1966.

Gibbs, R.J., Circulation in the Amazon River estuary and adjacent Atlantic Ocean, *J. Mar. Res.*, **28** (2), 113–119, 1970.

Katz, E.J., Further study of a front in the Sargasso Sea, *Tellus*, **21** (2), 259–269, 1969.

Kiyomitsu, K., A note on the salinity structure of the E. Bering Sea (in Japanese: English Abst.). *Bull. Tohoku Reg. Fish. Res. Lab.*, **30**, 79–85, 1970.

Knauss, J.A., An observation of an oceanic front, *Tellus*, **9** (2), 235–237, 1957.

LaFond, E.C. and K.G.LaFond, Studies of oceanic circulation in the Bay of Bengal, *Bull. Natl. Inst. Sciences, India*, **38**, 164–183, 1968.

LaViolette, P.E., Temperature, salinity and density of the world's seas: Bay of Bengal and Andaman Sea, *Naval Ocean. Off., Inf. Rept.*, No.67–57, Washington, D.C., 1967.

Perry, R.B. and G.R.Schimke, Large-amplitude internal waves observed off the north-west coast of Sumatra, *J. Geophys. Res.*, **70** (10), 2319–2324, 1965.

Uda, M., Researches on "siome" or current rip in the seas and oceans, *Geophysical Magazine*, **11** (4), 307–372, 1938.

U.S. Navy Hydrographic Office, Summary of oceanographic conditions in the Indian Ocean, Spec. Publ. 53, 1960.

U.S. Navy Hydrographic Office, Atlas of surface currents, Indian Ocean, H.O. Publ., No.566, 1st ed. 1944, 1970.

Voorhis, A.D., The horizontal extent and persistence of thermal fronts in the Sargasso Sea, *Deep-Sea Res.*, Supplement to Vol. **16**, 331–337, 1969.

Voorhis, A.D. and J.B.Hersey, Oceanic thermal fronts in the Sargasso Sea, *J. Geophys. Res.*, **69** (18), 3809–3814, 1964.

Zaneveld, J.R., M.Andrade, and G.F.Beardsley, Jr., Measurements of optical properties at an oceanic front observed near the Galapagos Islands, *J. Geophys. Res.*, **74** (23), 5540–5541, 1969.

Problem Areas in Air-Sea Interaction

HANS ULRICH ROLL

German Hydrographic Institute
Hamburg

Abstract   On the occasion of the eightieth birthday of Georg Wüst a short survey is given on the present state of air–sea interaction studies in the different scales of oceanic and atmospheric motion in order to demonstrate the progress reached, since Wüst started such investigations some fifty years ago, and, on the other hand, to show the gaps and short-comings which still exist in our knowledge.

GEORG WÜST'S RELATIONS TO AIR-SEA INTERACTION

For several decades oceanographers and meteorologists used to follow differ-ent avenues of research when investigating the structure and behavior of their respective media. The former were mainly concerned with the processes in the oceanic depths, while the latter concentrated their interest on the upper layers of the atmosphere. The interface between ocean and atmosphere received but little attention. During the last two decades, however, the situ-ation has changed remarkably, air–sea interaction studies now being "in vogue" and rapidly increasing in number.

Among those few who very early recognized the importance of and applied themselves to the crucial exchange processes occurring at the sea surface was Wüst (1920) who, when investigating the evaporation from the ocean, under-took to measure the vertical gradients of temperature, water vapor pressure, and wind speed in the lowest atmospheric layers above the ocean. He already realized the disturbing effects on those profiles as caused by the presence of a ship if it is used as a carrier for the measuring devices. At a later date (Wüst, 1937) he found it necessary to discuss these problems in more detail and arrived at the result that special laws apparently govern the heat and moisture exchange very close to the sea surface. In order to promote clarifi-cation he encouraged further studies on this subject (Schröder *et al.*, 1939;

Roll, 1939; Bruch, 1940), which, unfortunately, had to be discontinued during the war.

Obviously, Wüst's feeling was not very far from our present interpretation. Thus, on the occasion of the eightieth birthday of this recognized oceanographer it seems worthwhile to shortly summarize in general terms how far we have got in the understanding and application of air–sea interaction since Wüst first pointed out the problem some fifty years ago.

SCALE EFFECTS IN AIR-SEA INTERACTION

Calculations of the annual heat-energy budget of the system of ocean and atmosphere (Malkus, 1962) have shown that there is an excess of radiative energy within the equatorial belt between 38° North and South, whereas poleward from these latitudes more energy is lost to space than is received from the sun. The annual balance is achieved through suitable circulations in the fluid portions of the earth, whereby the atmosphere is the primary carrier of energy from the tropics to the polar regions. Since the atmosphere absorbs only a small fraction of the incoming solar radiation, it is mostly heated from below, whereby the amount of latent heat, provided by the evaporating sea surface, vastly exceeds the amount of sensible heat. From all this it seems clear that the interchange of energy and matter holds a key position within the system of atmospheric and oceanic motions.

Table 1 Scales in Atmospheric and Oceanic Motions

Name	Characteristics	Dimensions	
		Time	Space (horizontal)
Microscale	Turbulent motion (molecular exchange at the sea surface)	Up to several minutes	Up to 100 meters
Convective Scale	Pronounced vertical motion	Several minutes to one hour	100 meters to 10 kilometers
Mesoscale	Tendency toward organized motion	Several hours	10 to 100 kilometers
Synoptic Scale	Cyclonic and anticyclonic vortices around vertical axes	Several days	100 to 1000 kilometers
Large Scale	Quasi-stationary circulations, planetary waves, anomalies of the "Großwetterlage", climatic fluctuations	Weeks to decades	1000 kilometers to whole ocean

Oceanographers and meteorologists generally agree that the ocean and atmosphere must be considered as a combined system. Together they react to external influences. Both media are intimately coupled through a number of highly complex exchange processes of energy and matter in which the different scales of motion are important. The ranges of these scales are not generally fixed, but for the purpose of this short discussion, let us assume the rather crude limitations and incomplete interpretations as given in Table 1.

The interaction mainly concerns the following quantities:

Energy (momentum, sensible heat, latent heat, radiative heat);
Matter (other than water vapor).

The whole process of air–sea interaction can perhaps be sub-divided into the following parts, (which are then discussed below),

(1) actual exchange at the sea surface,
(2) vertical distribution and release of the energy and/or matter gained by air-sea interaction in ocean and/or atmosphere,
(3) horizontal transport of the energy and/or matter gained by air-sea interaction in ocean and/or atmosphere.

(1) This certainly is a small-scale phenomenon characterized by molecular and turbulent diffusion processes. As far as sensible heat and water vapor are concerned it is generally accepted that their transfer between air and sea is effected in the first instant by purely molecular conduction and diffusion taking place in a very thin surface layer with prevailing laminar motion and strong vertical gradients, which extends over substantial parts of the sea. Over and below it are transitional layers where the laminar flow is gradually changing into turbulent flow.

Regarding the momentum transfer from the air to the sea the relevant processes are rather complex, due to the fact that surface waves and currents are generated through wind action and cause certain feed-back effects on the atmospheric flow. It seems to be clear, however, that molecular viscosity does not play a decisive part, if any, since the momentum exchange is brought about by pressure forces acting on the roughness elements of the sea surface. The complicated mechanism of the frictional coupling between ocean and atmosphere is not yet fully understood.

(2) The distribution in the vertical and the release of energy transferred by the small-scale processes of part (1) is mainly executed by motions of the convective scale and the mesoscale. From cloud photographs taken by satellites as well as from tracer studies in the upper mixed layer of the ocean we know that in these ranges there is a pronounced tendency for a certain regularity, for organized motion, convective cells and roll vortices being the main features of it. It appears that such an organized exchange will always

have a greater effect than a completely turbulent one, provided these organizations are sufficiently stable.

When studying air–sea interaction in the convective and mesoscale range we should determine the basic requirements for such organized motions as well as the conditions for their stability or the reasons for their fluctuation. Further, we should investigate whether there exist direct relations between these regular flow patterns in atmosphere and ocean.

(3) The horizontal transport of energy and matter exchanged between ocean and atmosphere is achieved by motions of the synoptic scale and of the large scale. The main features of the synoptic scale are cyclonic and anticyclonic vortices around nearly vertical axes whereas the large-scale motions are characterized by quasi-stationary circulation systems and anomalies from mean flow.

When considering air–sea interaction in these scales we first have to study the feasibility of synoptic- and large-scale computations of energy exchange components between air and sea as well as their variability both in time and space. Further it must be clarified which role these diabatic fluxes between air and sea play in the development and mechanism of synoptic- and large-scale systems both in ocean and atmosphere. Is it such that an improvement of understanding and prediction can be expected from taking into account those diabatic transfers?

When discussing air–sea interaction in the different scales of motion we should not fail to notice the following important aspect: According to theory (Kolmogorov, 1941) there may exist an inertial sub-range in the turbulence spectra where energy is transferred from larger eddies to smaller ones. This "cascade process" ends in dissipation of the energy by molecular viscosity, where the smallest disturbances in the flow finally convert their kinetic energy into heat. Having in mind that, in the atmospheric mesoscale range, part of the available potential energy generated in "hot towers" may be transformed into synoptic- or large-scale kinetic energy we realize that such flow of energy is in a direction opposite to the one postulated by turbulence theory (Reiter, 1969). At the same time some of the kinetic energy generated by convective motion is dissipated into smaller eddies and finally into heat of the environment in agreement with the cascade concept. Thus, in the mesoscale range of the atmosphere a distinct separation as regards the energy transfer processes takes place, the atmospheric mesoscale system being characterized by energy transfers to smaller as well as to larger eddy sizes. This mechanism deserves further study. The mesoscale range, which is not covered by present observational networks, seems to be of fundamental importance for our understanding of the energy dissipation processes in the atmosphere. Similar phenomena seem to exist in the ocean.

PRESENT SITUATION

Describing the present state of knowledge of air–sea interaction in the different scales of motion is a task that cannot be fulfilled completely and objectively in a few words. We have to restrict ourselves to the major issues. Therefore, the following is, to a certain degree, biased by the constraint of attempting to be both as short and concise as possible.

Microscale

It is within the microscale where the actual interchange between ocean and atmosphere occurs, thus this range of motion deserves, and has so far received, special attention. Sometimes air–sea interaction problems were even identified with the problems of the marine boundary layer, which is an inadequate limitation.

The results obtained until now are impressive. We succeeded in measuring the turbulent vertical fluxes of momentum, sensible heat, and also water vapor by applying the eddy correlation technique, which seems to be the most direct and most adequate approach.

Some light has been thrown on the structure and peculiarities of the highly stratified interface layer, less than 1 mm thick, which separates ocean and atmosphere. Not yet understood, however, is the mechanism of frictional coupling between air and sea, in particular the feed-back effects of the wavy surface on the air flow.

Moreover, our knowledge is generally restricted to sea conditions with light and moderate winds, which is deplorable because it is to be expected that the interchange will be more intense and more important with high wind speeds. This is particularly true with regard to the effect of sea spray on evaporation and on the momentum transfer. Here some welcome assistance has recently been provided by air chemistry. Measurements of tritium concentrations in the wall clouds of tropical hurricanes revealed the amount of water vapor originating from evaporation at the sea surface (Östlund, 1968).

As regards the radiative processes in the system ocean-atmosphere our knowledge is gradually increasing as measurements of incoming and outgoing radiation as well as of the radiation balance have become available in growing numbers. In spite of this, the coverage of the oceans by radiation measurements still is far behind that for other meteorological and oceanographic quantities.

Convective scale and mesoscale

Owing to observational difficulties these scales have so far not obtained the attention they really deserve. In the atmosphere pioneer studies made by aircraft or by using smoke plumes and even sea gulls as indicators provided some very useful information on cloud dynamics, cloud organization, and also on organized motion in the sub-cloud layer above the sea. Some evidence for a relationship between such convective cells and roll vortices in the atmosphere and sea-surface properties has been given.

Later on, pictures taken from satellites revealed the surprisingly high degree of mesoscale organization of convective cloudiness above the ocean. Outbreaks of cold air over warmer water are characterized by such regular circulations which have diameters between 20 and 100 kilometers and obviously develop when an adequate thickness of the mixed ground layer of the atmosphere is reached.

In spite of such information our knowledge on the dynamics of those atmospheric processes is still rather incomplete. For instance, there does not seem to exist any well established explanation for the fact, to be found in many satellite pictures, that two types of cellular organization occur, the "open-cell type" and the "closed-cell type". In an "open cell" polygonal clear areas are surrounded by cloud walls, which indicates a downward motion at the cell center and an updraft at the cell margin. The opposite occurs in the case of the "closed cells". Hubert (1966) offered the explanation that the direction of circulation in mesoscale cells might be determined by the vertical variation of eddy viscosity. His suggestion, which is based on the analogy to laboratory studies, is entirely hypothetical and still has to be verified in nature. If true, it might be shown that sea-surface properties exert an important influence on such type of atmospheric motion.

What is apparently lacking and seems to be difficult to achieve is the coordination between satellite pictures and adequate measuring effort at the air–sea interface and in the relevant atmospheric layers. Results obtained during Bomex (Barbados Oceanographic and Meteorological Experiment 1969) will hopefully fill this gap.

As regards the ocean, tracer studies have indicated that organized motions similar to those in the atmosphere exist in the upper mixed layer above the thermocline. These so-called Langmuir circulations, which consist of convective cells and/or roll vortices, even might constitute a typical and essential feature of this layer. The possible causes for such mixing processes are thermal convection owing to evaporative cooling at the sea surface and/or the shear flow instability of the Ekman boundary layer circulation in the surface layer of the ocean. There is some indication that the second cause is the main one (Faller, 1963, 1964; Faller and Woodcock, 1964).

Thus semi-organized circulation cells may be the characteristic mesoscale motions in the atmosphere as well as in the ocean. They must play an important role in transporting the properties exchanged at the air–sea interface up to higher atmospheric levels as well as down to the thermocline in the ocean. The conditions for their development and stability or instability are only poorly known. Some pieces of observational evidence seem to support a model wherein large atmospheric cells act upon the smaller Langmuir circulations in the ocean. Much more research is needed here.

Synoptic scale

Obviously this is the scale that is most closely connected to practical day-to-day forecasting of atmospheric and oceanic phenomena. Regarding the feasibility of synoptic computations of energy-exchange components between air and sea we are facing severe difficulties.

The following approaches offer themselves in this respect:

(a) direct computations of air–sea energy transfer by applying the transfer formulas that are based on local measurements;
(b) budget studies of a closed area or along an air trajectory (line integral).

(a) As the direct measurement of local air–sea fluxes by means of the co-variance technique is not yet applicable (and probably will never be) on a routine basis, we have to consider simpler formulas for the calculation of these transfers which only contain quantities furnished by the present observational network, e.g. sea-surface characteristics and representative atmospheric properties as wind speed, temperature, and moisture content measured at some level above the sea. When doing so we have to parameterize the physics of the exchange processes by introducing into the transfer formulas suitable coefficients which, so far, have not been established universally beyond any doubt. In addition, sampling errors may occur as most of those diabatic transports have a pronounced short-term variability (diurnal and interdiurnal). Moreover, those transfer formulas fail to provide a very important information, namely where the energy taken up at the sea surface is released in the atmosphere.

(b) Consequently, integral methods for estimating the energy taken up by the atmosphere or lost by it seem to be more adequate, when synoptic-scale systems are considered, than putting together point measurements of air–sea interchange. Such quantities, however, are difficult to assess on a routine basis.

The next problem to be discussed is the role played by diabatic air–sea fluxes in the development and mechanism of synoptic-scale systems. Here the situation is different for such motions in the atmosphere and in the ocean.

In the atmosphere this problem has not yet been settled. As regards tropical disturbances, it is generally accepted that the fluxes of sensible heat and, particularly, of latent heat coming from the sea contribute significantly to the development of these vortices. As far as the depressions in moderate latitudes are concerned, the potential importance of the diabatic sea–air transports for cyclogenesis still has to be shown. While it is not surprising that the inclusion of the fluxes of sensible and latent heat appears to have a marked effect on the temperature pattern of a cyclone, mostly in the cold air mass to the rear of it (Pettersen *et al.*, 1962), it is difficult to see how this energy might contribute to the development of the depression. During cyclogenesis the diabatic fluxes seem to be second-order effects compared with baroclinic effects (Spar, 1965). Therefore, potential–into-kinetic-energy conversion still remains to be the dominant energy transformation in cyclogenesis. Perhaps, the heat transfer from sea to air helps to generate baroclinity in the atmosphere and thus contributes to the subsequent cyclogenesis, but a cyclone, once it is formed, is relatively unaffected by such diabatic fluxes.

In the ocean problems are perhaps a bit more settled. The wind stress is the most important one of the manifold influences the atmosphere exerts on the ocean. In the synoptic scale this effect is strongest in the wake of a hurricane where a temperature decrease up to $7\,^\circ$C was found at the sea surface due to the upwelling of cold water.

In spite of the fact that the frictional coupling between atmosphere and ocean is not fully understood, the theory of wind-induced oceanic motions is well advanced even for the case of non-stationary motions in a stratified ocean (Schmitz, 1964, among others). The difficulty is that so far it has not been possible to verify these theoretical results, which look very reasonable, by adequate measurements. The situation is more favorable with such effects as wind setup and storm surges. Here hydrodynamic-numerical methods have been successful in hindcasting sea-level variations which broadly agree with the relevant measurements at coastal stations.

In summary with regard to the effect the ocean exerts on atmospheric motions of synoptic scale, neither the observational evidence nor theoretical control are satisfactory at present.

Looking at the oceanic motions of synoptic scale which are caused by atmospheric influences we see that the situation is somewhat more advanced as far as the theoretical side is concerned, whereas the observational varification suffers from similar handicaps as in the atmosphere. For this reason it seems to be too early to give pertinent advice to the relevant international and national agencies regarding the setting-up of a world-wide ocean station network, in particular as far as its space and time characteristics are concerned. What is urgently needed is pilot studies under controlled conditions, as was the case with ATEX (Atlantic Trade Wind Experiment carried out by the research vessels *Meteor* (FRG), *Planet* (FRG), *Discoverer* (USA) and

Hydra (U.K.) in 1969) and BOMEX (1969), in order to clarify the role of the diabatic air-sea fluxes in the development and mechanism of synoptic-scale systems both in ocean and atmosphere and to determine how to assess these quantities for the purpose of synoptic predictions.

Large-scale

Large-scale motions in ocean and atmosphere comprise a wide range of temporal variations from weeks up to several decades. Suitable quantities for study are the departures of actual monthly or annual averages from the corresponding long-term climatological mean values. It is of particular importance to look for and, if possible, to establish quantitative relations between such anomalies in atmosphere and ocean which would serve as useful tools for the long-term forecasting of atmospheric and oceanic motions.

At first the methods employed were mostly descriptive and statistical. Later, there were endeavours to take into account the complicated physics of the actions and reactions of both media. Bjerknes and Namias are the pioneers in this field, the former demonstrating associations between anomalous sea-surface temperatures and the corresponding circulation for more or less decadal periods, while the latter mostly dealt with monthly or seasonal anomalies of sea-surface temperature and of atmospheric properties.

A rather advanced numerical model for predicting monthly ocean temperatures (Adem, 1969) is based on the equations of conservation of thermal energy both in the troposphere and in the surface of the earth. The equations include the storage of energy in ocean and atmosphere, the horizontal heat transport in the troposphere, the excess of radiation in the troposphere and at the surface of the earth, the vertical fluxes of sensible and latent heat from the earth's surface to the troposphere as well as the heat released in the troposphere by condensation. Although the horizontal heat advection in the ocean is neglected, the month-to-month changes in ocean-temperature anomalies as predicted by the model were, regarding their sign, correct 62.7% of 26 cases, which is distinctly higher than the result reached by the return-to-normal method (58.3%). An improved model will also include the horizontal heat transport by ocean currents.

Therefore the numerical methods have been successfully introduced into the calculation of large-scale interactions between ocean and atmosphere, and further progress in this field is to be expected. Naturally, when attempting long-term forecasting for years and decades, we must not consider only the sea-surface temperature as sufficiently representative for oceanic effects, but we have to take into account the exchange between the shallow mixed surface layer and the deep ocean which might have a significant effect on secular changes in atmosphere and ocean.

CONCLUDING REMARKS

When looking back over fifty years of air-sea interaction studies we realize a remarkable increase in the information about and understanding of the coupling between ocean and atmosphere. This development shows a marked acceleration, since the fastest progress occurred during the last two decades, and there is a strong indication that this tendency will continue.

Nevertheless, we are still far away from having acquired profound knowledge on how the two fluid media of the earth interact in the different scales of motion. We possess pieces of information confined to processes in single scales, but we know very little how the energy exchanged at the interface air-sea, chiefly by molecular conduction and diffusion, is transferred to other parts of the fluctuation spectrum. In particular, it should be explained why the direction of such energy transport is, in certain cases, opposite to the classical "energy cascade concept" of turbulence theory which says that the spectral energy decreases with growing wave number "k", i.e. that the energy transport is toward the smaller-scale motions. The desired explanation probably lies in the fact that three-dimensional isotropic turbulence, which is the necessary pre-requisite for the classical energy cascade, is very seldom realized in atmospheric and oceanic motions. In the atmosphere, the vertical transport of properties exchanged at the air–sea interface is mainly achieved, at least in the tropical region, by strong convective motions of the mesoscale with a locally pronounced preference given to the vertical flow component or to regular circulations in a vertical plane. Later, the horizontal flow components are preponderant. Consequently, the motion cannot be considered as three-dimensional isotropic but must approximately be taken as two-dimensional isotropic. Similar conditions exist in the ocean which is strongly stratified with regard to density and, therefore, favors horizontal motions. The theory of two-dimensional isotropic turbulence implies that the energy cascade is toward lower wave numbers if the classical wave-number relationship ($k^{-5/3}$) applies. This would be a "backwards cascade", energy being transferred from small-scale toward larger-scale motions.

Certainly, there is much evidence in meteorology and oceanography that energy can be transported out of turbulence and into the mean flow, but much more research is needed to clarify this important point beyond any doubt so that we may arrive at a better understanding of the way how the combined system of ocean and atmosphere really works in the different scales of motion.

References

Adem, J., Numerical prediction of mean monthly ocean temperatures, *J. of Geophys. Res.*, **74** (4), 1104–1108, 1969.

Bruch, H., Die vertikale Verteilung von Windgeschwindigkeit und Temperatur in den untersten Metern über der Wasseroberfläche, Veröffentl. Inst. f. Meereskunde, Berlin, N.F., Reihe A, No. 38, 1940.

Faller, A. J., An experimental study of the instability of the laminar Ekman boundary layer, *J. of Fluid Mech.*, **15**, 560–576, 1963.

Faller, A. J., The angle of windrows in the ocean, *Tellus*, **16**, 368–370, 1964.

Faller, A. J. and A. H. Woodcock, The spacing of windrows of Sargassum in the ocean, *J. of Mar. Res.*, **15**, 22–29, 1964.

Hubert, L. F., Mesoscale cellular convection, Meteor. Satellite Lab., Report No. 37, 1966.

Kolmogorov, A. N., The local structure of turbulence in an incompressible viscous fluid for very large Reynolds numbers, *C.R. Ac. Sciences*, USSR, **30** (4), 299, 1941.

Malkus, J. S., Large-scale interactions. In *The Sea: Ideas and Observations* (M. N. Hill, ed.), **1**, 88–294, Wiley (Interscience), New York 1962.

Östlund, H. G., Hurricane tritium II: Air-sea exchange of water in Betsy 1965, *Tellus*, **20** (4), 577–594, 1968.

Pettersen, S., D. L. Bradbury, and K. Pedersen, The Norwegian cyclone models in relation to heat and cold sources, *Geophys.*, Publ. Geophys. Norwegica, **42**, 243–280, 1962.

Reiter, E. R., Atmospheric transport processes, part 1: Energy transfers and transformations, U.S. Atomic Energy Commission, Division of Technical Information, 1969.

Roll, H. U., Zur Frage des täglichen Temperaturganges und des Wärmeaustausches in den unteren Luftschichten über dem Meere, *Arch. Seewarte u. Mar. Obs.*, **59**, No. 9, 1939.

Schmitz, H. P., Modellrechnungen zu winderzeugten Bewegungen in einem Meer mit Sprungschicht, *Deutsch. Hydr. Zeitschr.*, **17**, 201–232, 1964.

Schröder, B., H. U. Roll, and G. Seifert, Bericht über die meteorologischen Arbeiten während der Internationalen Golfstromuntersuchung 1938 auf Dampfer Altair, Abhandl. Preuss. Akad. Wiss., Phys.-Math. Kl., No. 5, Anlage 1, 1939.

Spar, J., Air-sea exchange as a factor in synoptic-scale meteorology in middle latitudes, ESSA Technical Note 9 – SAIL – 1, 1–16, 1965.

Wüst, G., Die Verdunstung auf dem Meere, Veröffentl. Inst. f. Meereskunde, Berlin, N.F., Reihe A, No. 6, 1–96, 1920.

Wüst, G., Temperatur- und Dampfdruckgefälle in den untersten Metern über der Meeresoberfläche, *Meteor. Zeitschr.*, **54**, 4–9, 1937.

The Hydrodynamic Roughness of the Sea Surface

K. BROCKS AND L. KRÜGERMEYER

Meteorologisches Institut Universität Hamburg
Hamburg, Germany

Abstract The effects of the vertical air stratification on the wind profile within the maritime boundary layer are investigated. Under unstable conditions the KEYPS-formula is used, under stable conditions the "log-linear"-profile. If the wind profiles under non-neutral conditions are approximated by logarithmic height functions, apparent roughness parameters "z_0" and apparent friction velocities "u_*" are obtained. Thus, variations of the roughness parameter of more than eight orders of magnitude are received.

By means of more than 1000 profiles with simultaneous measured wind, temperature, and humidity, from the Baltic Sea (1958) and in the North Sea (1959), values of apparent roughness parameters are derived and plotted as a function of the wind speed. The dependence of the Richardson number—related to the geometric mean of the height interval 1–10 m—on the wind speed is also examined. This results in a widely scattered distribution of "z_0" which systematically depends on the Richardson number. So far, this scattering was attributed to large scale effects of the state of the sea (fetch and duration). Under careful consideration of the stability, i.e. by using the virtual Richardson number, 152 profiles for near-neutral conditions $(\mathrm{Ri}_{v3.17m}) < 0.01$ are selected. The mean value of the drag coefficients—reference level 10 m—amounts to $C_D = 1, 3 \cdot 10^{-3}$. The values obtained by profile measurements in the Baltic-, the North-Sea and in the equatorial Atlantic Ocean practically coincide. The results of the profile measurements are also consistent with direct measurements of the vertical momentum flux (obtained by hot wire-, thrust-, and sonic anemometers). The discrepancies between previous measurements are due to disturbing influences during observation and due to neglecting the stability. With the aid of the mean wind speed the parametrization of the wind friction within the maritime boundary layer appears to be possible. Further investigations of the influence of the stability are intended.

INTRODUCTION

Air-sea interaction research requires a detailed knowledge of the vertical momentum and energy fluxes within the sea surface layer. Their experimental determination is, however, complex. The practical problem, therefore, is to relate these quantities to meteorological parameters obtained from routine

75

measurements over sea. This "parametrization", i.e. the coupling of the vertical fluxes with mean values of wind speed, air temperature, and air humidity at a given height as well as with the water temperature, requires an integration between the sea surface and the observation height involving information on the hydrodynamic properties of the boundary surface. Thus, the equation for the shearing stress τ at the sea surface

$$\tau = \varrho C_D (u - u_w)^2 \qquad (1)$$

ϱ air density
u wind speed at the height of observation
u_w current velocity at the sea surface, mostly neglected against u
C_D drag coefficient with respect to the observation height

makes it possible to determine the wind friction. Similar relations may be derived for the vertical fluxes of heat H and latent heat E.

Equation (1) follows from the semi-empiric theory of atmospheric boundary layer turbulence. In addition to the assumptions on the hydrodynamic properties of the sea surface and the diffusive processes within the surface layer, the vertical air density stratification ("stability") plays a major part in this theory. For a "hydrodynamic rough" sea surface with the roughness parameter z_0 the vertical wind shear may be expressed (e.g. Lumley and Panofsky, 1964) as:

$$\partial u/\partial z = (u_*/k)\, \phi\, (z/L)/z \qquad (2)$$

or, in integrated form, neglecting the height dependence of the shear stress:

$$u = (u_*/k)\, [\ln (z/z_0) - \psi\, (z/L)] \qquad (3)$$

$$C_D = k^2\, [\ln (z/z_0) - \psi\, (z/L)]^{-2} \qquad (4)$$

The symbols are defined as:

k v. Karman constant ($k \sim 0,4$)
u_* friction velocity

$$L = -(u_*^3 c_p \varrho \theta)/gHk \qquad (5)$$

where L is the stability parameter due to Monin-Obukhov with the dimension of a length,

c_p specific heat at constant pressure
H vertical heat flux
g gravitational acceleration
θ potential air temperature
$\phi\, (z/L)$ stability function,

from which the function $\psi(z/L)$ is derived by integration. Under neutral conditions there is

$$\phi(z/L) = 1$$

and

$$\psi(z/L) = 0.$$

From equations (2) and (5) the relation between the Richardson number

$$\text{Ri} = g\,(\partial\theta/\partial z)/T\,(\partial u/\partial z)^2 \tag{6}$$

and the dimensionless stability parameter z/L is obtained as:

$$z/L = (K_h/K_m)\,\text{Ri}\,\phi\,(z/L) \tag{6a}$$

where K_m and K_h are the eddy diffusivities for momentum and heat.

The vertical water vapor stratification is—particularly over sea—of considerable influence on the vertical density stratification. This may be taken into account by introducing a virtual potential temperature (e.g. Jeske, 1965):

$$\theta_v = (1 + 0.604q)\,\theta$$

where q means the specific air humidity. Thus, the virtual Richardson number Ri_v and the stability length L_v are obtained.

In order to determine the drag coefficient both the roughness parameter and the stability function $\phi(z/L_v)$ must be known. Both have to be determined experimentally.

The above mentioned equations are based on the assumption that the hydrodynamic roughness of the sea surface does not depend on vertical stability. It must be mentioned, however, that this assumption is not valid necessarily. Observations (Seilkopf, 1939) indicate that the appearance of the sea surface—its "morphology"—depends on the temperature difference water–air. Under stable conditions (air warmer than water), the sea surface appears to be "smoother" than under unstable conditions, the wind speed at 10 m height being the same in both cases. The stability has a similar influence on the "Beaufort-equivalents" of the wind velocity (Roll, 1954). But an influence on the hydrodynamic roughness could not be stated so far.

DETERMINATION OF THE ROUGHNESS PARAMETER BY PROFILE MEASUREMENTS

Under neutral conditions the following relations between friction velocity, roughness parameter and the wind profile may be derived from equations (2) and (3):

$$u_* = k\,\partial u/\partial \ln z \tag{7}$$

$$\ln z_0 = \ln z - [(1/u)\,\partial u/\partial \ln z]^{-1} \tag{8}$$

The drag coefficient is then given by

$$C_D = k^2 \, [\ln (z/z_0)]^{-2} \tag{9}$$

Provided the requirement for neutral density stratification is exactly fulfilled and the equations (2) and (3) are applicable, the quantities on the left
side of the equations (7–9) may be obtained from wind profile measurements.
Usually, the wind profile (i.e. measurements of the wind speed at different
heights up to 10 m) is fit by means of the method of the least squares in a
"$u - \ln z$-coordinate system":

$$u = a \ln z + b \tag{9a}$$

Then, the quantities of interest are given by

$$u_* = ak$$

and

$$\ln z_0 = -b/a$$

In the non-neutral case, however, the following relations between vertical
wind shear, friction velocity and the roughness parameter are obtained from
equations (2) and (3):

$$u_* = k \, (\partial u/\partial \ln z) \, [\phi \, (z/L_v)]^{-1} \tag{10}$$

$$\ln z_0 = \ln z - [(1/u) \, \partial u/\partial \ln z]^{-1} \, \phi \, (z/L_v) - \psi \, (z/L_v) \tag{11}$$

Thus, without the knowledge of the stability function these quantities—as
well as the drag coefficient—(see equation 4) cannot be determined by means
of the wind profile method. The application of the equations (7–9) would
involve considerable errors; in this case an "apparent" friction velocity "u_*"
and an "apparent" roughness parameter "z_0" would be obtained (Lettau,
1949). The relations between the "apparent" and the true quantities are
derived by combining the equations (10, 11) and (7, 8):

$$\text{"}u_*\text{"} = u_* \phi \, (z/L_v) \tag{12}$$

$$\ln \text{"}z_0\text{"} = \ln z - [\ln z/z_0 - \psi \, (z/L_v)] \, [\phi \, (z/L_v)]^{-1} \tag{13}$$

In order to demonstrate the influence of the stability on these quantities
analytic functions were substituted for the stability function in equations
(12, 13). In case of unstable conditions the so-called "Keyps"-profile
(Panofsky, 1963) was used, and for stable conditions the "log-linear"-
profile (e.g. Webb, 1970), i.e., for unstable conditions:

$$[\phi \, (z/L_v')]^4 - \gamma \, (K_h/K_m) \, (z/L_v') \, [\phi \, (z/L_v')]^3 = 1$$

$$\phi \, (z/L_v') = [1 - \gamma \, (K_h/K_m) \, \mathrm{Ri}_v]^{-1/4} \tag{14}$$

where the function $\psi\,(z/L_v)$ was derived from table 2 of Yamamoto's paper (Yamamoto, 1959), or, for stable conditions:

$$\phi\,(z/L'_v) = 1 + \alpha\,(K_h/K_m)\,(z/L'_v)$$

$$\phi\,(z/L'_v) = [1 - \alpha\,(K_h/K_m)\,\mathrm{Ri}_v]^{-1} \qquad (15)$$

$$\psi\,(z/L'_v) = 1 - \phi\,(z/L'_v)$$

L'_v is defined by $L'_v = (K_h/K_m)\,L_v$, and the ratio of the eddy diffusivities K_h/K_m is assumed to be constant. The following values were used:

$$\gamma K_h/K_m = 18; \quad \alpha K_h/K_m = 4.5; \quad z_0 = 1.46 \cdot 10^{-2}\,\text{cm} \quad \text{(see below)}.$$

Table 1 reveals the dependence of the "apparent" friction velocity on stability. The respective dependence of the "apparent" roughness parameter is represented in Figure 1. The logarithm of "z_0"/z_0 is shown as a function of the inverse stability length $1/L'_v$, or of the Richardson number with respect to a height of 3.17 m; this was done for the levels 3.17 cm, 31.7 cm and 3.17 m, i.e. for the geometric means of the height intervals 1–10 cm, 0.1–1 m, and 1–10 m, respectively.

Table 1   "u_*"/u_* = f (Ri). Ri < 0 "KEYPS"–Formula. Ri > 0 Log–Lin Profile (see text)

Ri	−1	−0.5	−0.1	−0.05	−0.01
"u_*"/u_*	0.48	0.56	0.77	0.85	0.97

Ri	+0.2	+0.15	+0.1	+0.05	+0.01
"u_*"/u_*	10	3.1	1.8	1.29	1.05

From Figure 1 it is evident that the value of the roughness parameter is falsified considerably, if the wind profile under non-neutral conditions is approximated at a given height by its tangent—representing a logarithmic height function—and if equation (8) is used. Similar errors arise if the wind profile is fit over height intervals, the geometric means of which correspond to the given heights.

Profile measurements over the ocean are carried out in most cases between 1 and 10 m. If the Richardson number is related to the geometric mean of this interval, its variations between −0.20 and +0.10 cause variations of the "apparent" roughness of more than three orders of magnitude. Even in the lower height intervals deviations from the true value z_0 may not be neglected.

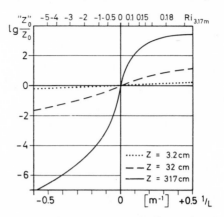

Figure 1   The apparent roughness parameter "z_0" as a function of $1/L$ and of $Ri_{3.17m}$. The "z_0"-values are obtained by the tangents to the profiles at the heights of $z = 0.032$ m, 0.32 m and 3.17. For unstable conditions the KEYPS-profile $(\gamma \cdot K_h/K_m = 18)$ was used and for stable conditions the log-linear profile $(\varkappa \cdot K_h/K_m = 4.5)$. $z_0 = 1.46 \cdot 10^{-2}$ cm

Only simultaneous measurements of wind-, temperature- and humidity-profiles permit the determination of the virtual Richardson number to such an extent that a decision on the applicability of the equations (7–9) is possible.

Several previous investigations of the hydrodynamic roughness of the sea surface did certainly not consider this point sufficiently. It is, therefore, to presume (see also Brocks, 1955), that some discrepancies between values of these quantities (Roll, 1965) may be attributed to this fact.

An instructive example for this is the paper of Kitaigorodskii and Volkov (1965). In Figure 3 of their paper the roughness parameter is shown as a function of the friction velocity. This representation results in a practically arbitrary distribution of more than 1000 values between 10^{-6} and 10 cm, similar to our Figure 1. As the authors point out "we need not follow for the influence of stratification" (quotation from the English translation), most of their points presumably represent "apparent" values of "z_0" and "u_*". Further conclusions cannot be drawn from this representation of "z_0" and "u_*". The opinion of the above cited authors that the relation between z_0 and u_* is "random" distributed and the conclusion of Kraus (1968) that "the profile method is for the birds" are not sufficiently justified with regard to this kind of material.

If adequate observations of the vertical density stratification are lacking, it is tried occasionally to get information on the stability indirectly via the shape of the wind profile. This method is based on the assumption that in the "ln $z - u$-coordinate system" a straight line as the best fit indicates neutral stratification. These conclusions, however, have to be drawn also very carefully, see Figure 2. Figure 2 shows a wind profile for neutral air together with three wind profiles for non-neutral conditions again obtained with the aid of the values of Yamamoto (1959) and characterized by the virtual Richardson numbers $Ri_{v,3.17m} = -0.18$, -0.69 and -3.4, and $\gamma K_h/K_m = 18$. The friction velocity is 0.41 m/sec and the roughness parameter amounts to $1.46 \cdot 10^{-2}$ cm. Wind velocities are figures for five different heights being characteristic for wind profile measurements over sea.

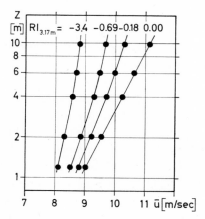

Figure 2 Windprofiles due to the KEYPS-formula (Yamamoto 1959) under different stability conditions, i.e. $Ri_{3.17m} = 0$, -0.18, -0.69 and -3.4. $z_0 = 1.46 \times 10^{-2}$ cm, $k = 0.41$, $u_* = 41$ cm/sec, $\gamma \cdot K_h/K_m = 18$

The main effect of the stability is to cause a different shape of the profiles. The curvature, however, is under normal conditions very small. Therefore, it is extremely difficult to decide on the strength of the profile shape whether the stratification is "neutral" or "non-neutral". It seems that. when this method was used, the "apparent" values of the roughness parameter and the friction velocity were often regarded as "true" values.

EXPERIMENTAL DIFFICULTIES

The "parametrization" method makes it possible to investigate turbulent transfer processes within the maritime boundary layer. Since the relation between the wind field and the state of the sea is neither over shallow water nor near the coast representative of deep water conditions, the required measurements have to be made over the open sea.

The vertical profiles of wind speed, air temperature and water vapor are sensitive to external influences. This means that measurements from ships or racks are not accurate enough. It is, therefore, necessary to use special buoys or floating platforms far from the ship's influence which do not affect the measurements and are not sensitive to the motion of the sea, thus not modifying the profiles. During recent years such measuring techniques have been applied more frequently. This is obviously the reason that more recently conducted research on the roughness of the sea surface and the drag coefficient shows a better consistency of the values than earlier investigations do (Roll, 1965). Evidently, a good deal of the discrepancies was attributed to unfavourable measuring conditions.

THE PROFILE MEASUREMENTS IN THE BALTIC- AND NORTH SEA 1958/59

Since 1957 the Meteorological Institute of the University of Hamburg has been working on air–sea interaction research with special emphasis on developing suitable measuring techniques (Brocks, 1959; Dunckel, 1967; Brocks and Hasse, 1969). Numerous campaigns have been made in the Baltic- and North Sea, and in 1965 (Brocks, 1966) and 1969 (Brocks, 1968) also in the tropical Atlantic Ocean. Firstly, only profiles of wind speed, air temperature and water vapor have been measured. Later, with the aid of a gyro-stabilized buoy the vertical fluxes of momentum and heat were directly measured by using the cross-correlation method (Hasse, 1968).

In the following, profile measurements made during three voyages in the Baltic Sea (Kiel Bight, spring and fall 1958) and in the North Sea (German Bight, fall 1959) are discussed. Some results have already been published (Deacon and Webb, 1962; Brocks, 1963; Roll, 1965). The present study gives information on the hydrodynamic roughness of the sea surface and the drag coefficient under neutral conditions.

The instruments (Lambrecht type cup-anemometers and aspiration psychrometers with platinum resistance elements) were mounted on a mast carried by a buoy. The buoy was floating at a distance of some hundred meters from the research vessel "Hermann Wattenberg" belonging to the "Institute of Oceanography of the University Kiel".* Stabilized by a weight the buoy and the mast showed only small tilt motions even in heavier sea (Augstein and Wucknitz, 1969). More than 1000 profiles (averaged over 15 min periods) were obtained. The wind speed was recorded by a device giving a pulse per every 100 m distance covered by the wind. The water-,

---

* We wish to express our warmest thanks to Prof. Dr. Wüst for making his ship available to us.

of the Atlantic Expedition 1965—vertical profiles of wind, temperature, and humidity were measured by means of an improved meteorological buoy (Dunckel, 1967). From 787 profiles obtained under near-neutral conditions ($|\text{Ri}_v\,(4\text{ m})| < 0.01$) being representative for the south-east trades Hoeber derived the following values:

$$C_D = (1.23 \pm 0.25)\,10^{-3}$$

resp.

$$z_0 = 1.1 \cdot 10^{-2}\text{ cm}.$$

The agreement is remarkable and so is the relatively large scattering of the individual 15 min averages.

By means of Table 2, taken from Hasse (1968) and subsequently supplemented by a more recently obtained measuring series, a comparison with direct measurements of the vertical momentum flux is possible. The series 1, 2 and 4 were, however, obtained with the aid of masts over shallow water near the coast, and the series 3 was obtained by means of "mast measurements" in the Baltic Sea.

Using direct "buoy measurements" carried out in the Mediterranean by Zubkovski and Kravschenko (1967), Hasse (1968) derived the following mean value:

$$C_D = (1.83 \pm 0.84)\,10^{-3}$$

Hasse, following Stewart, explains the unusual large standard deviation as well as the observed increase of the cross-correlation $r\,(u',\,w')$ with increasing u by the remaining pitching of the buoy. So we don't wish to consider these measurements within this comparison either.

These results obtained by different measuring methods demonstrate that more recent measurements show almost no systematic differences (see also the statements of Miyake *et al.*).

Table 2   Direct measurements of the drag coefficient

Author	Meanvalue $C_D(10\text{ m}) \cdot 10^3$"standard deviation"
1 *Smith* (1967)	1.03 ± 0.18*
2 *Weiler and Burling* (1967)	1.31 ± 0.36
3 *Hasse* (1968)	1.21 ± 0.24
4 *Miyake et al.* (1970)	1.09 ± 0.18 Reynold stress measurements
	1.13 ± 0.18 profile measurements

* Due to Hasse (1968) only values obtained by a properly adjusted instrument were averaged.

The remaining discrepancies are much less compared to the so far discussed scattering of the roughness parameter of more than eight orders of magnitude. Undoubtedly, the better consistency is due to the new measuring techniques (use of special meteorological buoys) and to the careful consideration of the density stratification.

The drag coefficient for the North Sea and Atlantic Ocean profiles shows practically no wind dependence, the respective dependence for the Baltic Sea measurements is somewhat more distinct, though small. The "best fit" for all measurements in 1958/59 (Figure 5) is given by

$$C_D \cdot 10^3 = 1.18 \pm 0.18 + 0.016 u_{10}$$

In Figure 6 the mean values (and the "standard-deviations") of this collective are plotted against the wind speed subdivided in sections. Also shown are the Atlantic measurements. Accordingly, the corrective factor of the above equation appears to be unimportant, and for many purposes the above mentioned mean value will suffice.

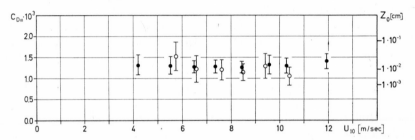

Figure 6   Mean values and standard deviations of the drag coefficient as a function of the mean wind speed (at 10 m) derived from wind profiles under neutral conditions. The stability was obtained from simultaneous wind-, temperature- and humidity profiles. $|\text{Ri}_{v3.26\,\text{m}}| < 0.01$. Dots: Baltic- (1958) and North Sea (1959) measurements; Circles: Equatorial Atlantic (1965) measurements

The following values were derived by Deacon (1962):

$$C_D \cdot 10^3 = 1.10 + 0.04 u_{10},$$

and by Sheppard (1958):

$$C_D \cdot 10^3 = 0.80 + 0.114 u_{10}.$$

As to these relations one has to consider that a major part of Deacon's data were obtained from ship measurements and from mast measurements over the mud flats near the coast. Sheppard relies on five profile series (wind profiles only) which were measured by Hay (1955) by means of a platform. The platform nearly 800 m off a 60–90 m high cliff was moored at a depth of 25 m. A fetch of 800–1200 m was observed. It appears that open sea conditions were, thus, not met by this.

Wu (1968) relying on values of the drag coefficients published by different authors derived two interpolation-formulas. He did, however, not check the respective measuring series critically. For a reference level of 10 m Wu derives the following relation:

$$C_D \cdot 10^3 = 0.5u^{0.5}, \quad 1.0 < u < 15.0 \text{ m/sec.}$$

This relation is plotted in Figure 5. As may be seen, this curve shows a too large increase of the drag coefficient with increasing wind speed compared to our measuring values. Analyzing wind profile data Charnock (1955) came to the conclusion that "z_0 was comparatively independent of fetch and stability but was largely determined by u_*".

He suggested as the "simplest non-dimensional relation":

$$z_0 = u_*^2/ag \tag{16}$$

where a amounts to 80. Kitaigorodskii and Volkov (1965) determined this constant from their widely scattered data to $a = 28.5$. Using these a-values the relation (16) is plotted in Figure 5 too. Both curves fit only a part of our values. The best agreement is achieved for a value of $a = 50$. By this the systematic difference (i.e. the too large increase of C_D with increasing u_{10}) is maintained.

It is worth mentioning, however, that the "z_0"-values are rather well approximated by Charnock's relations (16) under unstable conditions, see Figure 3. Similarly, Kitaigorodskii and Volkov will have achieved the respective consistency of their mean values with equation (16). The authors, however, call special attention to the widely scattered values which due to their opinion may not be attributed to the local mean wind field (see below).

We suggest that the agreement between the Charnock-relation and wind profile observations indirectly includes the effects of the density stratification under unstable conditions prevailing over the ocean.

A decrease of "u_*" and, respectively, a stronger decrease of "z_0" is caused by unstable stratification. These reduced values of "z_0" (see Figure 3) are roughly approximated by equation (16). This is because for small values of "u_*" the effect of the density stratification is much greater.

Kitaigorodskii and Volkov (1965) assumed "that the roughness projections move in the direction of flow at some velocity c" and derived the following relation for the roughness parameter:

$$z_0 \propto H \exp\left(-kc/u_*\right) \tag{17}$$

In equation (17) H means the height of the wind effected sea. For c the formula for the phase velocity of waves with the period P was used: $c = gP/2\pi$. The authors applied equation (17) to wind profile measurements and to measurements of the state of the sea. In this way they are able to represent the values of z_0—derived from the wind profile method—within a range of 6 orders of magnitude. This result is remarkable in so far, as these values presumably represent apparent roughness parameters "z_0" and apparent friction velocities "u_*".

Unfortunately, we are not able to make a corresponding comparison since the wave periods have not been measured during the Baltic and North Sea voyages. But due to the range of wind speed measurements (2.5–14 m/sec) and due to the different sea areas considerable differences of the quantities characterizing the state of the sea have to be expected.

Occasionally a parametrization of the wind friction by means of the local wind velocity has been doubted. It has been pointed out that the properties of the state of the sea being responsible for the wind friction are governed by large scale parameters as fetch, duration and by surface tension rather than by local processes. It was further mentioned that non-linear effects— independent of the local wind speed—might cause a shifting within the wave spectrum which also might influence the wind friction. The observed differences of the roughness parameter of several orders of magnitude were regarded as an indication for this.

When discussing this problem in future, one should consider that stratification effects might simulate a considerable variability of the roughness parameter and the friction velocity. Our results show, however, that even if considering the stratification influence strictly, a scattering of the values of the roughness parameter cannot be avoided. This remaining scattering is fortunately much smaller if considering average values and drag coefficients. It is not clear whether the above mentioned additional influences are the cause for this remaining scattering.

This is not necessarily to be explained by variations of the large scale parameters (fetch and duration). It seems to be conceivable that this scattering is characteristic for the fluctuations of the vertical wind shear which are presumably random distributed within a turbulent field and therefore are reduced with regard to large area means. If this assumption is correct, the averaging of different profiles shown in Figure 6, would yield values of the drag coefficient, which if combined with the local wind speed due to equation (1) give the effective wind friction averaged over the area under consideration.

The agreement of recent profile measurements with direct measurements of the vertical flux of momentum seems to indicate the correctness of this assumption. Thus, the parametrization by means of the local wind speed

ment between June and October, these zones stretch from 80° W to 170° E in the Pacific, and thereby cover a total of about 150° of longitude. Thus they are large enough to control the planetary average (Figure 1). Under these conditions, it is possible for the "Intertropical Convergence Zone" (ITCZ) to split above the Pacific and Atlantic, which occurs regularly over Africa (Strüning and Flohn, 1969) and the Indian Ocean.

The occurrence of the equatorial countercurrent in the ocean presents some similarities (and dissimilarities) with the atmospheric processes. Stimulated by Prof. Wüst's lectures in Bonn from 1964 to 1967, the Meteorological Institute has produced several contributions to this very fascinating and many-sided problem. While, in spite of the convincing charts of water temperature anomalies in all oceanographic atlases, the conception of an equatorial cold source (Fletcher, 1945; Flohn, 1949) has not been further investigated. The author has investigated an extensive series of ship observations along the Europe–South America route in order to produce a detailed structural analysis (Flohn, 1957). Unfortunately, the approach taken was incomplete: in order to assure incontestable divergence and vorticity calculations in a 1° × 1° grid, the evaluation had to be semi-annual to incorporate enough data (366,000 observations). D. Henning, at the suggestion of H. Riehl, has now evaluated similar data from the Atlantic ship routes to South America and South Africa on a monthly basis (Henning, 1970). He has found that along the South African route especially, the role of the equatorial cold water is clearly evident. M. Hantel has contributed an investigation of the surface wind divergence over the Indian Ocean (Hantel, 1970a).

Besides the calculation of average conditions, we are also interested in temporal fluctuations. These are especially interesting because they are accompanied by unusual and economically important fluctuations of rainfall amount in the area of the Atlantic and Pacific equatorial arid regions and along the arid west coasts of Africa and South America between about 2° and 16° S. After a preliminary study of possible Trans-Atlantic correlation of temporal fluctuations (Eickermann and Flohn, 1962) Doberitz (1969 applied modern statistical methods which he first used in an investigation o the equatorial Pacific area (Doberitz *et al.*, 1967; Doberitz, 1968a, 1968b to study the Atlantic, but could only partly confirm the earlier results of Eickermann and Flohn. Especially interesting, however, is his finding of a negative correlation between the anomalies of the South American Pacific coast (and thereby the entire Pacific equatorial zone) and those of the equatorial Atlantic, i.e., a sort of anomaly see saw.

Usually the remarkable equatorial cold water zone is interpreted, after J. Bjerknes (1966), as the result of a divergence of the Ekman drift of the water surface at both sides of the equator, and therefore finally as a result of a sign difference of the Coriolis parameter. This two-sided divergence, which can also exist in a weakly convergent wind field, produces a rising of the sub-surface cool water or at least of the lower mixing zone near the thermocline, in the same way as the one-sided divergences on the well-known cold water coasts. If J. Bjerknes' theory is correct, then the maximum development of this phenomenon should be reached at a time when the wind field is symmetrical with regard to the equator. This is the case, for example, in northern winter (January–February) when the (simple) Intertropical Convergence Zone lies approximately over the equator and the trade winds of both hemispheres are of about equal intensity. Strangely, however, this pheno-menon occurs in the Atlantic and Eastern Pacific mainly in the northern summer (June–October) when the ITC is 10–15° from the equator and the SE-trades cross the equator broadside with remarkably high speed. In the Gulf of Guinea SSE–SSW winds prevail on both sides of the equator, and in the Pacific ESE–SE winds prevail; in both cases the atmospheric stream-lines turn clockwise (Figure 2). These systems are also accompanied by a divergence of the Ekman Drift.

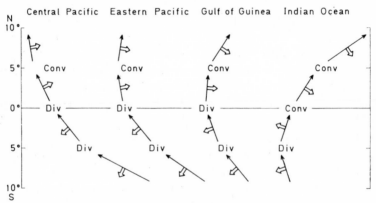

Wind (⟶) and Ekman Drift (⇑) in Equatorial Latitudes, N – Summer.

Figure 2   Average wind vector and vector of the Ekman Drift (vertically integrated up to the frictional depth)

The sign of the meridional component D_Y of the vertically integrated Ekman-Drift current (ϱ = density):

$$D_Y = -(\varrho f)^{-1} \tau_x$$

(where $f = 2\Omega \sin \phi$ is the coriolis parameter) is dependent on the zonal component τ_x of the average surface stress vector τ, whose direction is very nearly the same as that of the resultant wind vector, and on the sign of f

which changes at the equator. Because $f \to 0$ at the equator, this area cannot be included. Figure 2 shows schematically the arrangement of divergence and convergence of D by different typical wind distributions in the area of the equator.

The divergence of the wind field, as calculated from observations along the two ship routes from Europe to South America and from Europe to South Africa show, according to Henning (1970), a remarkable isolated equatorial maximum in the months June to October and centralized in July (Figure 3) at 2° N. In the same months, the temperature difference between water and air (Figure 4) shows an isolated zone at 1–5° S where $T_L - T_W > 0$ centralized at 3° S. The vorticity of the wind field shows, according to Flohn (1957) and Henning (1970), a wide extension of anti-cyclonic vorticity on the northern hemisphere in northern summer; the line curl$_z$ $\tau = 0$ coincides approximately with the ITC, and the highest value lies at 6° N. According to Hantel (1970b) the vertical velocity at the lower boundary of the Ekman layer is, in accordance with $w = (\varrho f)^{-1}$ curl$_z$ τ, dependent on the vorticity of the surface stress vector τ and the Coriolis parameter. The distribution of curl$_z$ τ is qualitatively very similar to that of curl$_z$ V (Hantel, 1970c).

CROSS-SECTION E-ATLANTIC (0-20°W), JULY

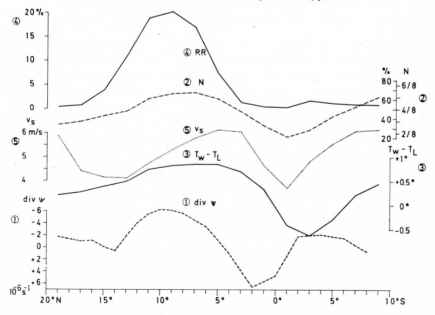

Figure 3 Meridional behavior of meteorological parameters in the Atlantic, July. Data of D. Henning (1970) where T_w = sea surface temperature, T_2 = air temperature, div VI = divergence of the wind field, RR = frequency of rainfall in percent of all observations, N = apparent cloud cover, V_s = average scalar wind velocity

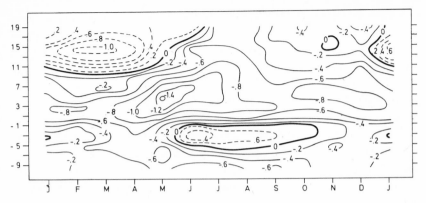

Figure 4   Isopleths of the temperature difference between water and air along the
Europe–South Africa ship route (after D. Henning, 1970)

In this connection we should also mention the correlations between the
anomalies of water temperature T_W and rain amount RR (monthly values
after removal of the average annual cycle) for Canton Island (2.8° S, 171.8° W,
184 months, 1950–65) and Fernando de Noronha (3.8° S, 32.4° W, 233 months,
1906–13 and 1922–38) according to an as yet unpublished report by R. Doberitz. While the data from Canton Island are from the station itself, the T_W
data from Fernando de Noronha are taken from observations published by
Bullig (1954) from an approximately 5500 km² ocean area (3–5° S). The
persistence of T_W anomalies is remarkably great: only after 5–6 months do
the autocorrelation coefficients sink under 0.50, and after 9–11 months under
the level of significance. The correlation of simultaneous monthly anomalies
of T_W and RR is for Canton Island +0.56 and for Fernando de Noronha
+0.30. Both of these coefficients have a probable error of ±0.04 and lie well
over the 3σ significance level of 0.22 or 0.20. Over half of the covariance is
produced by significant periods of >15 months. The phase difference between T_W and RR is 0.3 months with the rain before the temperature. As far
as this unexpected result can be considered representative, it points to an
anomaly caused by the atmosphere (possibly from the wind field) rather than
a simple stabilization effect of the cold water.

A consideration of the distribution of the temperature difference sea–air
(Figure 4) leads us to some considerations about the heat balance. In the
months July–October, the flow of sensible heat (U_L) in the zone 1–5° S is
directed unmistakably downwards $(T_L > T_W)$. Because there are no humidity
data available, it is impossible to give any exact figures for the vapor pressure
difference $e_W - e_L$. However, since no fog over the equatorial cold-water
areas has been observed (in contrast to the wide-spread fog areas off the
coasts of Angola, SW-Africa and Peru), it seems safe to assume that the flow

On the Response of a Stratified Deep Ocean to Wind and Air Pressure

WOLFGANG KRAUSS

Institut für Meereskunde, Kiel University
Kiel, Germany

Abstract Response functions for horizontal velocity components and vertical displacements due to variable wind and air pressure are given. Wave numbers cover the range from 10^{-7} cycles to 10^{-2} cycles/m, frequencies range from 10^{-4} cycles/hour to 10^2 cycles/hour.

The energy is mainly transmitted from the atmosphere to ocean along the dispersion lines of internal, inertial and surface waves.

INTRODUCTION

The existence of a large variety of periodic, quasiperiodic and random fluctuations has been apparent since the earliest measurements of velocity and temperature from anchored ships. Evidently the fluctuations occur whether or not strong currents exist. During the first four decades of this century, A. Defant, V. W. Ekman and many others finally concluded that inertial waves, internal waves, and eddies most likely dominate these fluctuations, which seem to be superposed on well known barotropic modes, e.g., tidal currents.

Extensive current measurements during recent years in the Baltic Sea (W. Krauss, 1966, 1968) and in the Atlantic Ocean (F. Webster, 1968a, 1968b) made possible the computation of spectra of current fluctuations over frequency ranges from ca. 10^{-2} to ca. 10 cycles per hour. These spectra, in general, show a peak at the inertial frequency with a strong decrease in energy with increasing frequency and a particularly slow increase with decreasing frequency. Another characteristic of the spectra is that different slopes occur at frequencies above and below 10^{-2} cycles per hour. Amplitudes at tidal frequencies might vary considerably.

Measurements in both the Atlantic Ocean and the Baltic Sea indicate strong time and depth variation in currents. Coherence between adjacent stations is, in general, low and poorly understood at the present time. Inertial oscillations are easily interpretable by Fredholm's solution of the drift current problem (V. W. Ekman, 1905, 1941). Because wind fields are not stationary in most areas of the oceans, wind-produced currents adjust toward mean Ekman currents in the form of oscillations with a period of a half pendulum day.

Fredholm's theory has been supplemented by the work of J. E. Fjeldstad (1958) on the adjustment of currents toward steady state, showing again, that any transition to a new balanced condition in mean currents is accompanied by inertial oscillations. Indeed, the fact is well known that inertial waves have been observed in most areas of the ocean (A. Defant, 1960), including adjacent seas like the Baltic Sea or even large lakes (J. L. Verber, 1964). In contrast to these theories and facts, many scientists adopted the concept that inertial waves are created by tidal waves at a critical latitude, i.e., where the tidal frequency equals the inertial frequency. F. Webster (1968a) gives a complete summary of these views. In addition, reference should be made to the early publications of V. W. Ekman and J. E. Fjeldstad; from their work, inertial waves as a typical response behaviour of an extended fluid on rotating earth can be deduced. Altogether, the concepts on inertial waves seem to approach slowly the old and more realistic picture of the thirties.

The role played by internal waves and eddies in the entire current spectrum is by no means clear at the present time. Whether or not large fluctuations in temperature and current velocities are due to eddies or internal waves has been an open question since the publication of the fundamental textbook on the Norwegian Sea by B. Helland-Hansen and F. Nansen (1909). The eddy concept was emphasized during the 1950's within the scope of homogeneous turbulence theory but only in the high frequency range do spectral measurements of temperature fluctuations agree well with that concept. The shallowness of the World Ocean seems to prevent a general application of this theory to geophysical problems.

On the other hand, internal wave theory had not been sufficient to interpret observed fluctuations and their vertical dependency by decomposition into less than five modes except in a few cases. Several years ago, an extensive effort by the author, to decompose observed spectra of current fluctuations in the Baltic Sea into spectra of the first six modes, was not successful. This seems to indicate that internal wave theory in its simplest form (neglect of the wind produced effects and their typical depth dependency) is applicable only within certain frequency bands. It seems unlikely that energy is concentrated in higher modes rather than lower ones. What is needed now is a full account of the fundamental solutions of the Fredholm type for the inhomogeneous ocean which can be done by including friction into internal wave theory. A first step in this direction has been made by M. Tomczak (1967).

RESPONSE FUNCTIONS DUE TO WIND AND AIR PRESSURE

The response of a stratified deep ocean to wind and air pressure may best be illustrated by means of response functions, which display the response of the ocean to external forces. Examples for a homogeneous ocean have been given by V. S. Belyaev (1967) and A. D. Yampolski (1966). The theory for a stratified ocean was published by W. Krauss (1970). The same theory has been applied to deep water. All computations have been carried out on the CDC 1604 A of the Navy Electronics Laboratory Center, San Diego, California. Plots were made for five depths, $z = nH/5$, where $n = 0, 1, 2, 3$ and 4 on a Calcomp Plotter. The response functions are displayed in a frequency–wave number plane. Each figure is based on 73 points on the frequency axis and

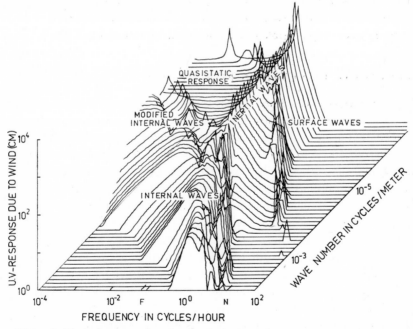

Figure 1 Typical response function for the horizontal velocity components u, v as function of frequency and wave number (the unit cm of the response function must be multiplied by the spectrum of τ/μ in order to yield cm sec^{-1}, τ being the tangential stress, μ viscosity)

45 points on the wave number axis. The frequency scale is logarithmic, the wave number scale, however, is a linear one.

In order to make the reader familiar with response functions in the frequency–wave number plane we refer to Figure 1, which is from W. Krauss (1970). The ocean responds to large-scale slowly varying meteorological fields in the form of quasipermanent currents and displacements. The response is higher only for certain resonance frequency–wave number combinations. These are:

Inertial waves occur for all wave numbers smaller than ca. 10^{-5} cycles per meter (wave lengths larger than ca. 100 km) at the inertial frequency.

Surface waves location is given by the characteristic equation for long waves, including friction. They show a branch point near the inertial frequency: one branch approaches the inertial waves and the other intersects them.

Internal waves position in the frequency–wave number plane is given by the characteristic equation for internal waves. Like surface waves they too show a branch point near the inertial frequency, one branch approximating the inertial frequency and the other intersecting the inertial range. They are referred to here as *modified internal waves* because their depth dependency differs from internal waves as sustained in a frictionless model. On a rotating earth they are only possible if friction is taken into consideration.

RESPONSE FUNCTIONS FOR DEEP WATER

The parameters selected are typical for the eastern part of the Norwegian Sea, where the most intensive studies on fluctuations of this type have been carried out during the last 60 years. The region may be described by: (see the list of symbols on p. 109) $H = 1000$ m, $\varphi = 63°$, Väisälä-period $= 48$ min, $\varkappa = \eta$, and $v_H = 10^{-2}k^{-1}$. Because of the strong mixing within the Norwegian current, $v_v = 500$ has been used mainly. Severe difficulties arose from exceeding the range for which exponential functions could be handled numerically. When this occurred, the response functions were truncated, as in Figure 2 (upper left corner) for frequencies beyond 10 cycles per hour.

The response functions for horizontal velocities u, v (u, v-Response) for five depths are shown in Figure 2. All response functions due to wind are dominated by inertial waves, and response functions due to pressure are dominated by very long surface waves. The response at the inertial frequency varies with depth but exceeds the neighbouring frequencies at all depths at least by $10–10^2$. The inertial frequency band, therefore, is the most significant zone in the ω, k-plane within which the energy of the wind fields can be transferred from the sea surface to great depths. For higher wave numbers the

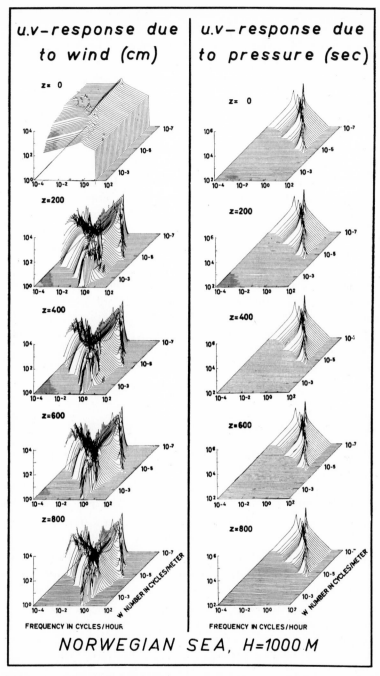

Figure 2 Response functions for the horizontal velocity components u, v (units with respect to wind induced currents as in fig. 1; pressure induced currents must be multiplied by the spectrum of air pressure gradient/density in order to yield velocity spectra)

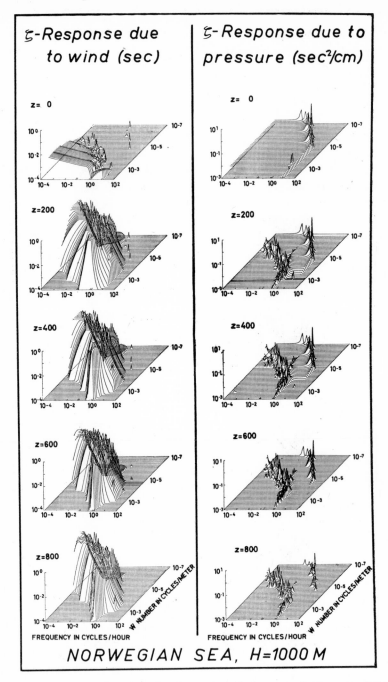

Figure 3 Response functions for the vertical displacement ζ (the unit of the response functions must be multiplied by $\tau/(\mu k)$ for wind induced displacement spectra and air pressure/density for air pressure induced ones ($k =$ wave number))

internal waves take over this role. The air pressure is of minor importance in deep water as an energy source for velocity fluctuations.

There is little doubt that the dominance of inertial waves at great depth in Figure 2 is partially due to the high vertical eddy viscosity $v_v = 500$. The computations, however, have been repeated on a linear scale including the inertial frequency for $v_v = 40$. The general features are essentially the same. In contrast to shallow waters, where only three modes prevail, about 10 modes are present in the internal wave branches, of the deep sea. From this we may conclude that at least 5–10 modes are necessary to describe the fluctuations adequately in the internal wave range.

The ζ-response functions due to wind stress and air pressure have also been evaluated for deep water (Figure 3). They are entirely dominated by internal waves. We conclude from these figures that most energy is transmitted into the deep ocean at the dispersion lines for internal, inertial and surface waves.

Symboles used in text and figures

H depth

v_H horizontal eddy coefficient

v_v vertical eddy coefficient

$k = (\varkappa^2 + \eta^2)^{1/2}$ horizontal wave number

ω frequency

ζ vertical displacement

φ latitude

References

Belgaev, V.S., The dependence of the spectra of the velocity components of a wind-driven current on the spectrum of the tangential wind force. Izvestiya Acad. Sci., USSR, *Atmospheric and Oceanographic Physics*, **3** (11), 1217, 1967.

Defant, A., *Physical Oceanography*, Vol. 1. Pergamon Press, 1960.

Ekman, V.W., On the influence of the earth's rotation on Ocean Currents. Arkiv Math., *Astron. och Fysik.*, Bd. 2, No. 11, 1905.

Ekman, V.W., Trägheitsschwingungen und Trägheitsperioden im Meer. *Ann. d. Hydrogr. u. Marit. Meteorol.*, **69**, 238–249, 52 pp., 1941.

Fjeldstad, J.E., Ocean currents as an initial problem. *Geofys. Publ.*, **20**, 7, 1–24, 1958.

Helland-Hansen, B. and F. Nansen, The Norwegian Sea. Kristiania 1909.

Krauss, W., Interne Wellen. Gebr. Borntraeger Verlag, Berlin, 250 pp., 1966.

Krauss, W., Typical features of internal wave spectra. *Progress in Oceanography*, Vol. 5, 95–102, Pergamon Press, 1968.

Krauss, W., On the response of a stratified ocean to wind and air pressure, *Dtsch. Hydrogr. Z.* (in press) 1970.

Tomczak, M., Jr., Über den Einfluß fluktuierender Windfelder auf ein stetig geschichtetes Meer. *Dtsch. Hydrogr. Z.*, **20**, 101–129, 1967.

Verber, J.L., The detection of rotary currents and internal waves in Lake Michigan. *Proc. 7th Conf. on Great Lake Res.*, Univ. Michigan. Great Lakes Research Div., Publ. No. 11, 382–389, 1964.

Webster, F., Observations of inertial-period motions in the deep sea. *Review Geophysics*, Vol. 6, No. 4, p. 447, 1968a.

Webster, F., On the representativeness of direct deep-sea current measurements. *Progress in Oceanography*, Vol. 5, 3–16, Pergamon Press, 1968b.

Yampolskii, A.D., The dependence of the velocity spectrum of drift currents on the spectrum of the tangential wind stress. Izvestiya Acad. Sci., USSR, *Atmospheric and Oceanographic Physics*, **2** (11), 1186, 1966.

Observation of Laminae in the Termocline
of the Tropical Atlantic*

FEODOR OSTAPOFF

National Oceanic and Atmospheric Administration
Atlantic Oceanographic and Meteorological Laboratories
Miami, Florida, 33130

Abstract   A 15-hour time series was obtained in the tropical Atlantic (10°30′ N and 43°20′ W). The sampling period was five minutes and the principle tool used was the Expendable Bathythermograph (XBT). The data is presented in form of depth variation of selected isotherms, of temperature variations at selected depth and of depth variation of two selected features in the vertical temperature profiles.

INTRODUCTION

During the Atlantic Tradewind Expedition (ATEX) last winter the USC&GS Ship *Discoverer* drifted for 18 days within the tradewinds and carried out a multidisciplinary research program. The primary objective of the expedition, in which four research ships from three nations participated (see Figure 1), was the investigation of air–sea interaction processes in the tropical Atlantic Ocean. The main program consisted of 8 meteorological balloon ascent per day sounding the atmosphere up to 7 kilometers height with simultaneous oceanographic soundings down to 500 meters. Eight expendable bathythermographs (XBT) were dropped daily to obtain temperature profiles to 450 meters. In this way it was hoped to record temperatures vs. depths at 90-minute intervals.

On February 16 a program was initiated which would record temperature profiles at 5-minute intervals for 15 hours in order to study the variability of temperature and internal waves with periods up to 3 hours. Simultaneously,

---

* Presented at the ICES Symposium on "Physical Variability in the North Atlantic" held at Dublin, Ireland in September, 1969.

a salinity-temperature-depth sensing unit (STD) was suspended at 40 meters depth for the first 5 hours, at 100 meters for the next 5 hours and at approximately 275 meters for the last 5 hours. The purpose of these STD measurements was to investigate the variations in temperature and salinity in the mixed layer (40 m), the most intense part of the thermocline, (100 m), and the region below the thermocline (275 m).

After the termination of this program the routine 3-hourly XBT and STD program was resumed until the end of the expedition on February 22, 1969. This paper will present only the preliminary results from the oceanographic data obtained during the 15 hours of recording on February 16–17 at 10° 30′ N and 43° 20′ W (Figure 1).

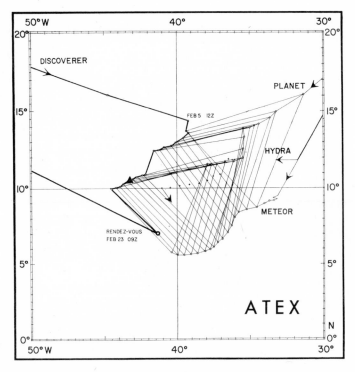

Figure 1   Solid triangle indicates the position at which the XBT Times Series was obtained during the Atlantic Tradewind Expedition 1969 (ATEX)

INSTRUMENTATION

The principle instrument used to gather the data for this study was the expendable bathythermograph (XBT) manufactured by Sippican Corporation. The XBT system is a thermistor embedded in the nose of a streamlined body that is released from a launcher on board ship. The probe falls through the water with a constant velocity of 7 m/sec. The time constant for the thermistor is $\frac{1}{10}$ sec. Trailing wire connection transmits the signal back to the recorder. After reaching design depth (in this case 450 meters) the wire breaks, and the recording ends. Although the probe is designed for underway operation with ship speeds up to 30 knots, in this study it was used to obtain vertical profiles at one location in rapid succession.

The other instrument used in this study was the STD, also called the bathysonde. The particular model (Bissett–Berman Model 9006) records temperature, salinity, sound velocity, and pressure as a function of time. According to the manufacturer, the time constant for the temperature probe is 0.35 seconds.

THE OBSERVATIONS

The basic characteristics of the vertical temperature distribution are shown in Figure 2. The surface layer is well mixed and extends to 63 meters in this profile. The mixed-layer depth during the entire period varied between 40 and 70 meters. The main thermocline exhibits a mean temperature gradient of 10 °C/60 m or 0.17 °C/m. However, the structure in the thermocline is such that layers with high gradients alternate with layers of either low gradients or constant temperature, resulting in a step-like temperature distribution with depth. Locally, the gradient reaches values of at least 1 °C/m and it may exceed the limits of our observational tool. Laminae or nearly uniform layers occur about at a rate of 5 per 50 meters in the upper part of the thermocline. Similar features have been observed by White (1967) in the thermocline, in much deeper layers by Stommel and Fedorov (1967) and on a smaller scale by Woods (1968).

Also shown in Figure 2 are two laminae, one around 75 meters in the upper thermocline (labeled 1) and another one most prominent near 150 m (labeled 2). This extensive laminar layer of up to 30 meters thickness separates the upper part of the thermocline from the lower part. This lower part has a mean temperature gradient of 1 °C/24 m or 0.042 °C/m. Again, the temperature distribution shows step-like features all the way down to the termination of the last 450 meters.

Figure 3 shows the first 13 complete temperature profiles out of some 180. Each profile was obtained at a 5-minute interval, hence the figure shows complete profiles for one hour. The continuity of the major laminar is obvious.

As mentioned earlier, the STD sensor was kept for a period at each of three depths. From inspection of the pressure record, it is seen that the

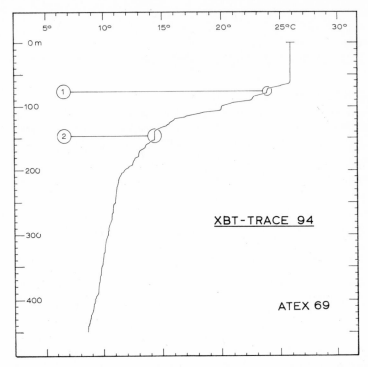

Figure 2 A typical XBT profile showing principle features of normal stratification on February 16, 1969 0315 L.T. at 10° 30′ N and 43° 20′ W. Features 1 and 2 are the two predominate laminae and are referred to in Figure 7

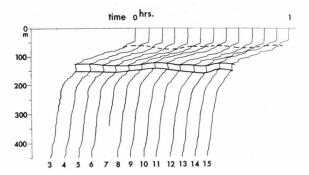

Figure 3 Reproduction of XBT traces for the first hour of the 15-hour time series. Mixed layer temperatures about 26 °C and 450 meter temperatures about 9 °C. Dashed line indicates feature 1 and solid lines show feature 2 in Figure 2

instrument moved up and down with an amplitude of about 2 meters and a period of about 7 seconds due, no doubt, to heaving of the ship. While the temperature trace shows at times a high "noise" level attributed to a strong vertical temperature gradient, at other times the record looks quiet on all channels except the pressure. The most likely explanation seems to be the movement of a lamina through this depth level. Obviously, in this case the temperature record will reflect no vertical motion resulting from internal waves, which would have much longer periods than seven seconds.

In order to investigate whether the XBT time series at the chosen 5-minute sampling period reproduces the major fluctuations, temperature values at 30-second intervals were read from the STD record after smoothing of the high frequency noise. The solid line in Figure 4 represents the temperature variations at 100 meters from the STD record. The circles are XBT 100 meter temperatures and the triangles represent XBT 95-meter temperatures. The width of the shaded area between these, therefore, shows the mean temperature interval between 95 and 100 meters. It is clearly seen that periods of strong gradients alternate with periods of weak gradients during which time a lamina is moving through that depth range. It is also seen that the agreement between the XBT and STD data is particularly good during periods of weak gradients (within 0.2 °C). In general, the STD curve lies within the hatched area and all major features are reproduced by the XBT data. This is quite remarkable, especially in view of the large temperature fluctuations of more than 2 °C in 5 to 10 minutes.

The remaining figures represent various ways to reduce the XBT data. First, the conventional procedure is applied, namely, to plot the depth distribution of individual isotherms. In Figure 5, depth variations of the 25°-, 20°-, and 15°-isotherm are presented. The thin curves show running one-hour means while the thick lines represent the mean depth of each isotherm. Amplitude variations of more than 10 meters in 20 minutes, sometimes more than 20 meters, occurred quite frequently and were more pronounced for the 15°-isotherm than the 25°-isotherm. Although all three isotherms behaved similarly as far as the trend is concerned, the coherence in the high frequency variations is rather low. A striking example is the downward departure of the 15° isotherm from the hourly mean by 24 meters at 11 hr 10 min whereas the 20°-isotherm as well as the 25°-isotherm coincide exactly with the hourly mean. Large oscillations of the 15°-isotherm near 12 hours are not all followed by the 25°-isotherm.

The results of a second method of reducing the original data are presented in Figure 6. Here temperature is plotted versus time for several individual depth levels (60, 80, 100, 120, and 200 meters). Thick horizontal lines represent the mean temperatures for the corresponding depth levels during

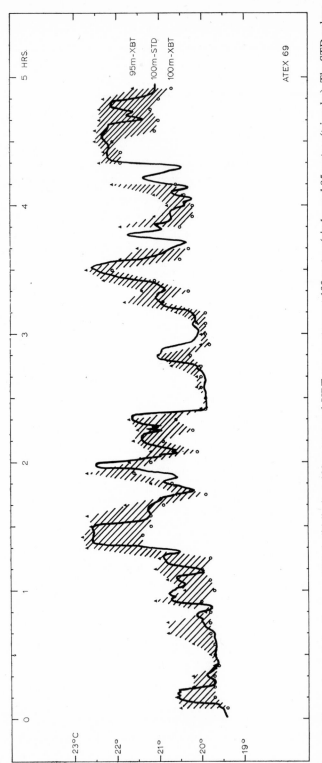

Figure 4 Comparison of simultaneous STD (solid line) at 100 meters and XBT temperatures at 100 meters (circles and 95 meters (triangles). The STD values were plotted at 30 second intervals

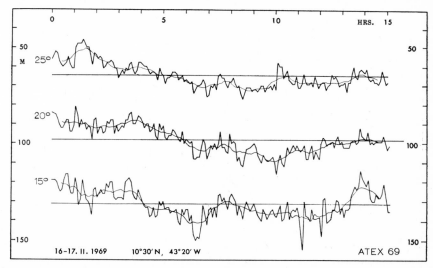

Figure 5 Depth variation of the 25°-, 20°- and 15°-isotherms for a period of 15 hours on February 16–17, 1969 at 10° 30′ N and 43° 20′ W

the 15 hour period. Their spacing represents the mean temperature gradient. As seen from the top curve, the surface mixed layer penetrated below 60 meters after 5 hours and remained for the rest of this time series. To some degree the deepening of the mixed layer is also reflected in the temperature record at 80 meters. The largest amplitude temperature variation is found at 100 and 120 meters with more than 5 °C amplitude. At and below 140 meters these changes do not exceed more than 2.5 °C. This is consistent with the mean temperature structure as shown in Figure 2. The small temperature variations at 140 meters can be attributed to the presence of a thick extensive lamina near this level seen in Figure 2.

Finally, the depth distributions of features 1 and 2 of Figure 2 have been followed as function of time. Figure 7 shows the vertical extent and depth of the 23° lamina and the 14° lamina. This should be the best evidence available for internal wave studies inasmuch as no specific assumption concerning the local temperature gradient is required in transforming temperature fluctuations at a given level into vertical motions. This record, more than any other, demonstrates little correlation between these two features. Moreover, at times the upper boundary and the lower boundary of either lamina vary with different amplitude and phase. The 23° lamina is growing thicker with time and, at the same time, cooling, while the 14° lamina is decaying and warming.

This discussion has been limited to a data presentation; the next step will consist of the detailed analysis of the various time series, their power spectra,

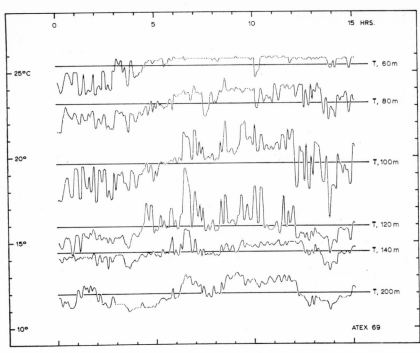

Figure 6 Temperature fluctuations at selected depth levels for a period of 15 hours on February 16–17, 1969 at 10° 30′ N and 43° 20′ W. The horizontal lines indicate the mean temperatures for the entire period at the respective depths

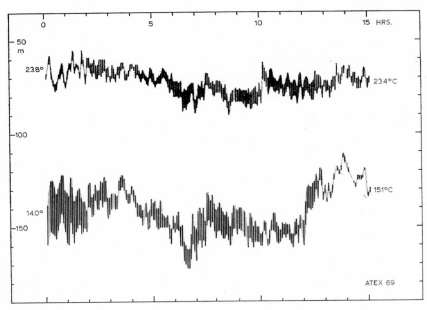

Figure 7 Thickness and depth variations of two selected features in the vertical temperature distribution (see Figure 2)

correlations, and coherences, which will be published elsewhere. Most interesting should be the comparison of the various time series in an effort to determine which represents the best approach to studying internal waves in the open ocean.

CONCLUSION AND SUMMARY

On the basis of a 15-hour time series of temperature profiles at 5-minute intervals the following conclusions can be made.

(1) The XBT is a useful tool to study the thermal structure of the thermocline in regions of strong vertical gradients.

(2) Large temperature variations at a given depth with amplitudes up to $5°C$ were found in the region of a strong thermocline.

(3) The vertical temperature structure in this region exhibits layering of sheets with strong gradients interrupted by layers of nearly uniform temperature. The temperature profile resembles a step-like function.

(4) If temperature records are obtained at preselected depths for studying internal waves, it is important to record not only the temperature fluctuations but also simultaneously temperature gradients. This is essential in regions with strong gradients and laminae.

References

White, R. A., The vertical structure of temperature fluctuations within an ocean thermocline. *Deep-Sea Res.*, **14**, pp. 613–623, 1967.

Stommel, H. and K. N. Fedorov, Small scale structure in temperature and salinity near Timor and Mindanao. *Tellus*, **19** (2), 1967.

Woods, J. D., Wave-induced shear instability in the summer thermocline. *J. Fluid Mech.*, **32** (4), pp. 791–800, 1968.

Paper 9

Wind Stress Curl–the Forcing Function
for Oceanic Motions

MICHAEL HANTEL

Meteorological Institute of the University of Bonn, Germany

Abstract The surface wind stress curl is the independent forcing function of models describing wind-driven ocean currents in terms of the vorticity equation. This study presents an annual and seasonal charts of the stress curl over the world ocean.

Basic data were the surface wind stress values computed by Hellerman (1967) for $5° \times 5°$-squares. The significance of the curl function for barotropic models is discussed.

The charts confirm the well-known latitudinal pattern of the curl—i.e., cyclonic curl over the polar caps, anticyclonic curl in middle and subtropical latitudes, cyclonic curl in the tropics—; however, they add some new details. Noteworthy deviations in the tropics are interpreted in connection with large-scale oscillations of the current systems, with climatic anomalies and with off-shore upwelling. In the tropical Indian Ocean, the periodically reversing curl field gives rise to the oscillating system of the monsoon currents. In the eastern parts of both the Pacific and Atlantic Ocean, the phenomenon of desert-like arid zones just south of the equator in northern summer is attributed to large-scale upwelling motion due to the negative curl areas in these regions.

1 INTRODUCTION

In his famous paper of 1950 Munk stated that "permanent ocean currents are related to the rotational component of the wind stress field over the ocean. These ocean currents would vanish were the wind stress field is rotational". The global pattern of the wind stress curl is the topic of the present contribution.

Following Stommel (1948) and Munk (1950), many authors have treated the wind-driven ocean circulation as a two-dimensional problem on the β-plane, thereby using the method of vertically integrating the hydrodynamic equations introduced by Ekman (1905) and Sverdrup (1947). For the sake of completeness, the various assumptions leading to the governing vorticity

121

equation

$$\frac{\partial}{\partial t} + J + \beta - A + k = \text{curl} \tag{1}$$

(short-hand notation for eq. A. 16)

are assembled in Appendix A. Eq. (1) is a short-hand notation of eq. (A.16), denoting the time-derivative term by $\partial/\partial t$, the nonlinear or inertial term by the Jacobian symbol J, the planetary vorticity term by the Rossby parameter β, the Navier–Stokes friction term by its coefficient A, the bottom friction term by its coefficient k, and the rotational part of the wind stress field by curl. Once the parameters of a barotropic nondivergent ocean model have been fixed, the left-hand side of eq. (1) depends on the (unknown) stream function ψ, whereas the right-hand side is the forcing function of the problem, determined not by the ocean's motion but by the atmosphere.

One of the reasons for the extensive literature about wind-driven ocean currents is the challenge of the Gulf Stream. As early as 1924 Wüst showed in his epoch-making paper "Florida- und Antillenstrom" three fundamental features of the Florida current: (1) The current is highly baroclinic, hence it allows the application of the dynamic (or geostrophic) method of Bjerknes to determine the velocity component perpendicular to an oceanographic section. This enabled Wüst, from simple hydrographic observations, to determine quantitatively the vertical velocity profile across the Florida current as well as its total water transport. (b) The current is due to the pressure gradient in the water, not to the local wind. (c) There is a countercurrent north of approximately 28° N flowing south.

It is interesting to note that these features are also exhibited by barotropic models based on eq. (1). The region of the western boundary current is in close geostrophic equilibrium in linear as well as in highly nonlinear models (Veronis, 1966). The local wind does not affect the boundary current, and the countercurrent is also a feature of both linear (Munk 1950) and nonlinear models. Veronis (1966) discusses in some detail the significance of such a two-dimensional homogeneous model for the motions of a three-dimensional inhomogeneous ocean.

In Table 1 a number of recent studies about barotropic wind-driven ocean circulations are listed. Some of them do not deal with special cases of the non-divergent vorticity equation (1), but rather with the divergent set of primitive equations (i.e., eqs. (A.12), (A.13), (A.7)); in these cases, the notation P.E. has been added to the author's name and the pertinent non-divergent vorticity equation in symbolic form has been added in order to allow for comparison. The range of the eddy viscosity coefficient A (or the bottom friction coefficient k) and of the Rossby parameter β has also been listed. The curl patterns used by different authors have been normalized for

Table 1 Recent baratropic wind-driven circulation studies. P. E. signifies that these papers deal with the divergent set of primitive equations

Author	Pertinent vorticity equation	Range of A [cm² sec⁻¹] or k [sec⁻¹]	Range of β [10^{-13} cm⁻¹ sec⁻¹]	$\mathrm{curl}_z\,\vec{\tau}$, proportional to	Special emphasis on
Munk 1950	$\beta - A = \mathrm{curl}$	$A = 5 \times 10^7$	1.2–2.3	$-\dfrac{\partial \tau_x}{\partial y}$, observed	General ocean circulation
Moore 1963	$J + \beta - A = \mathrm{curl}$	$A = 2 \times 10^7$	1.0	$-\dfrac{\partial \tau_x}{\partial y} \sim -\sin\dfrac{\pi y}{L}$	Nonlinear Rossby waves
Bryan 1963	$\dfrac{\partial}{\partial t} + J + \beta - A = \mathrm{curl}$	$A = 10^6\text{–}10^8$	1.0	$-\dfrac{\partial \tau_x}{\partial y} \sim -\sin\dfrac{\pi y}{L}$	Numerical Experiments
Fischer 1965 (P.E.)	$\dfrac{\partial}{\partial t} + J + \beta - A = \mathrm{curl}$	$A = 10^8,\,10^9$	1.5	$-\dfrac{\partial \tau_x}{\partial y} \sim \cos\dfrac{2\pi y}{L}$	Numerical scheme
Pedlovsky 1965	$\dfrac{\partial}{\partial t} + J + \beta + k = \mathrm{curl}$	–	–	$\cos(kx - \omega t)\sin n\pi y$	Time-dependent curl field
Veronis 1966	$\dfrac{\partial}{\partial t} + J + \beta + k = \mathrm{curl}$	$k = 1 - 4 \times 10^{-6}$	2.0	$\dfrac{\partial \tau_y}{\partial x} - \dfrac{\partial \tau_x}{\partial y} \sim -\sin\dfrac{\pi x}{L}\sin\dfrac{\pi y}{L}$	Comparison linear-intertial
Gates 1968 (P.E.)	$\dfrac{\partial}{\partial t} + J + \beta - A = \mathrm{curl}$	$A = 6 \times 10^7,\,10^8$	1.754	$-\dfrac{\partial \tau_x}{\partial y} \sim -\sin\dfrac{\pi y}{L}$	Transient Rossby waves
Hantel 1969 (P.E.)	$\dfrac{\partial}{\partial t} + \beta - A = \mathrm{curl}$	$A = 5 \times 10^7 - 5 \times 10^9$	2.3	$-\dfrac{\partial \tau_x}{\partial y} \sim -\sin\dfrac{\pi y}{L}$	Comparison numerical-analytical
Düing 1969	$\dfrac{\partial}{\partial t} + \beta + k = \mathrm{curl}$	$k = 10^{-9} - 10^{-4}$	0–2.48	$\dfrac{\partial \tau_y}{\partial x} - \dfrac{\partial \tau_x}{\partial y} \sim -\left(\sin\dfrac{\pi y}{L} + c\right)\sin \omega t$ $c = \text{const.}$	Forced oscillations, comparison friction-intertial regime

a quadratic ocean basin with sidelength L; the x-coordinate points eastward, the y-coordinate northward.

It is not our aim to dwell on the interrelationships and the significance of the various models listed in Table 1; this is done, to some extent, by Veronis (1966) and Gates (1968). We would rather like to make some comments about the role of the forcing function of eq. (1) which, obviously, is crucial for a proper solution of the vorticity equation—in particular when applied to the total world ocean.

As inspection of Table 1 reveals, Munk seems to be the only author who has based his treatment of the vorticity equation on the observed wind stress curl distribution over the ocean. Later on, interest was entirely confined to the ocean's response to idealized curl patterns. This approach was, in itself, a straightforward and a necessary one in order to learn about the mechanism of the fluid's motion (this viewpoint is particularly stressed by Veronis, 1963). However, investigation of transient flows requires studies over more extended regions (Veronis, 1966), and this implies some need for a closer look to the independent forcing function. Obviously, this remark applies also to the mechanism of stratified models (see, e.g., Veronis and Stommel, 1956, or Schmitz, 1964) which are, though, not considered in the present contribution.

2 GLOBAL PATTERNS OF SURFACE WIND STRESS CURL

There exist some recent compilations of the stress curl and of the wind vorticity which is closely related in pattern to the stress curl. Mintz and Dean (1952) have compiled charts of the surface wind vorticity for the world ocean for the months of January and July; Stommel (1965) has presented an annual mean chart of the surface stress curl. According to these authors, the north–south distribution of the vorticity and likewise of the curl is characterized by anticyclonic values in middle and subtropical latitudes and cyclonic values in the tropics. The general cyclonic vorticity in tropical latitudes is interrupted by a belt of anticyclonic vorticity just north of the equator in July, and just south of the equator in January.

These results are confirmed by the present evaluation which is based on the seasonal and annual wind stress data compiled by Hellerman (1967). For computational details see Appendix B. Figure 1 shows zonal averages of the wind stress curl over the oceans as a function of latitude. The unit is 10^{-8} kg/m^2/sec^2 which corresponds to 10^{-9} dynes/cm^3. Some features of the patterns of Figure 1 deserve special attention: (1) The stress curl pattern is fairly symmetric with respect to the equator, except in northern summer. (b) In the inner tropics there exists a strong annual oscillation. (c) The midlatitude anticyclonic curl belt in the southern hemisphere is broader, more intense and farther poleward than its equivalent in the northern hemisphere. This is in agreement with the well-known fact that the southern hemisphere's circulation is considerably stronger than the northern hemisphere's (see, e.g., Flohn, 1967a).

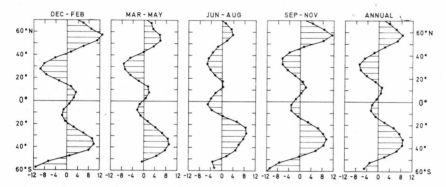

Figure 1   Zonal average of wind stress curl. Unit 10^{-8} kg/m^2/sec^2 = 10^{-9} dynes/cm^3

Figures 2 through 6 show the geographical distribution of the stress curl for the four seasons and the annual mean. Some of the important large-scale features are:

I   On the northern hemisphere, the curl is anticyclonic in the subtropics and cyclonic in middle and higher latitudes throughout the year but varies considerably in intensity. Strongest circulation appears in northern winter.

II   On the southern hemisphere, the curl is also anticyclonic in the subtropics and cyclonic in middle and higher latitudes. It varies in pattern but not in intensity. One can certainly conclude that the intensity of extratropical curl on the southern hemisphere is not stronger in southern winter than in southern summer. In particular, note the large-scale anticyclonic area southwest of southern Africa which is very strong and persistent throughout the year. In areas south of about 50°–60° S no definite statement can be made due to the lack of data.

III   Taking into account the generally smaller values of the curl in tropical latitudes, the seasonal fluctuations of the curl are remarkable over large areas of the tropics. In particular, note the patches of negative curl in both eastern Pacific and Atlantic which vary considerably in size and intensity throughout the year; we will return to this point in section 3. Likewise, note the reversal of the curl field in the Indian Ocean north of 20° S, a phenomenon which extends farther east into the western Pacific. An extensive discussion of the stress curl in the Indian Ocean based on data on a 2° × 2°-

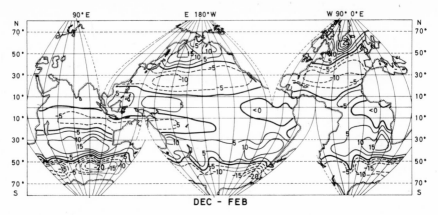

Figure 2 Surface wind stress curl for northern winter. Negative isolines dashed. Missing data areas indicated by dotted isolines. Same unit as in Figure 1

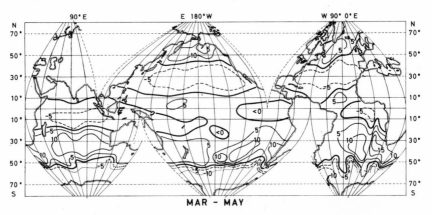

Figure 3 Surface wind stress curl for northern spring. Negative isolines dashed. Missing data areas indicated by dotted isolines. Same unit as in Figure 1

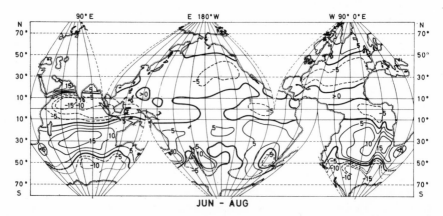

Figure 4 Surface wind stress curl for northern summer. Negative isolines dashed. Missing data areas indicated by dotted isolines. Same unit as in Figure 1

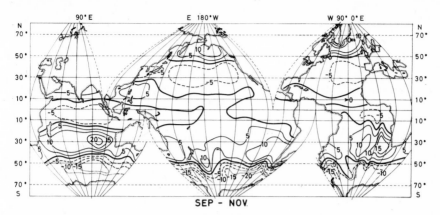

Figure 5 Surface wind stress curl for northern fall. Negative isolines dashed. Missing data areas indicated by dotted isolines. Same unit as in Figure 1

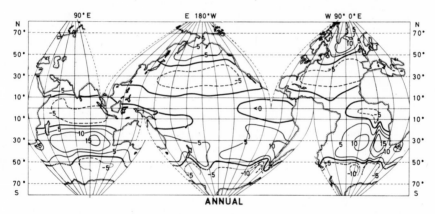

Figure 6 Surface wind stress curl, annual mean. Negative isolines dashed. Missing data areas indicated by dotted isolines. Same unit as in Figure 1

square and monthly scale is given elsewhere (Hantel, 1970). The curl fields in the cited study compare satisfactorily with the present ones over most of the Indian Ocean; however, discrepancies appear in the northern Arabian Sea and in the Bay of Bengal due to the coarser grid of the present data.

3 SOME APPLICATIONS

As has been outlined in section 1, the wind stress curl is the forcing function in models describing oceanic motions by a vertically integrated vorticity equation. This applies not only to barotropic or non-divergent models as described in Appendix A but also to layered and divergent models The charts presented here may serve as an observational background for further theoretical studies in this field.

In the Rossby Memorial Volume, Welander (1959) has presented a world map of the Sverdrup transport in the oceans based on the annual mean wind stress data of Hidaka (1958). The Sverdrup transport M_y is given by the following truncated form of eq. (A.16):

$$\beta M_y = \text{curl}_z \, \vec{\tau} \qquad (2)$$

We reject a repetition of this type of evaluation not only because of the truncations involved in (2) but also for another reason The time-lag between the forcing function and the barotropic circulation is of the order of 1 to 2 months (Düing, 1969; Hantel, 1969), the time-lag for baroclinic processes being considerably longer (Veronis and Stommel, 1956). As it is not yet clear to what extent the response of the actual ocean is barotropic or baroclinic, application of eq. (2) for varying wind stress curl, even on a seasonal scale, does not seem very meaningful. On the other hand, application of eq. (2), based on the annual mean field of Figure 6, would hardly improve Welander's Sverdrup transport map.

A well-known application of the curl field is concerned with off-shore upwelling. In the Hidaka Jubilee Volume, Stommel (1965) has presented, besides an annual chart of the wind stress curl (which compares sufficiently with Figure 6), a geographical table of the vertical velocity just beneath the Ekman layer. At the bottom of the mixed surface layer of the ocean the vertical velocity may be estimated (Yoshida and Mao, 1957) by

$$w = \frac{1}{\varrho f} \, \text{curl}_z \, \vec{\tau} \qquad (3)$$

If $w > 0$ the motion is directed upward. Eq. (3) balances the wind stress curl by the vertical velocity at the interface between the ocean's surface layer and the lower layers. Eq. (3) is, in a sense, complementary to eq. (2) which balances the stress curl by the meridional Sverdrup transport. The main difference in eqs. (2) and (3) lies in the application of the continuity equation. Eq. (2) is based on the assumption of non-divergence whereas, strictly speaking, the balance of eq. (3) is between the stress curl and the horizontal mass flux divergence in the surface layer.

If eq. (3) is valid for areas reasonably distant from the equator, say, for latitudes greater than 4°, we have to expect in the above-mentioned tropical regions of the eastern Pacific and eastern Atlantic with negative curl upwelling mainly on the southern hemisphere ($f < 0$) and in northern summer due to the greater intensity of the curl patches in this season.

It is, in fact, a well-known observation that intense negative temperature anomalies in the inner tropics are mainly confined to the southern hemisphere and that they are maximally developed in northern summer; this can be proven, e.g., by inspection of the sea surface temperature maps for the Pacific Ocean (Reid, 1969), and by inspection of similar maps for the tropical Atlantic Ocean (Mazeika, 1968). In these latitudes, vast areas of sea surface temperature, sometimes as low as 22.5°C (Atlantic, August), extend from the west coasts of southern America and Africa farther west into the middle of the oceans and spread out northward, partly entering the northern hemisphere. North of the equator, temperatures quickly rise in a narrow zone and reach the mean value for tropical seas, i.e., 27°C. This pattern is not confined to the northern summer season but is then strongest. The magnitude of the upwelling motion is, calculated from eq. (3) together with Figure 4, of the order 5×10^{-4} cm/sec or some 10 m/month, a value which compares reasonably with figures of Yoshida and Mao (1957), and others.

Climatologists have long been faced with the problem of understanding the excessive dryness of the areas of the eastern Pacific and eastern Atlantic just south of the equator, and the rainbelts of the ITCZ (Intertropical Convergence Zone) region just north of the equator. It is known that the ITCZ almost never penetrates into the southern hemisphere; this has been attributed to the asymmetry of the tropospheric circulation in both hemispheres (Flohn, 1967b). There are hints from preliminary sea–air interaction models that sea surface temperature is crucial for the ITCZ formation; related numerical experiments have shown that low temperature areas are forbidden regions for the ITCZ and hence give rise to negative anomalies of rainfall (Pike, 1970). If in anomalous winters the sea surface temperatures fail to be low extensive precipitation occurs in these desert-like south-hemispheric belts (in particular: Galapagos Archipelago, Peruvian coast) leading to the El Niño phenomenon (Doberitz, 1968). This phenomenon appears only in northern winter when the tropical negative curl patches are small and weak anyway.

From these considerations one is tempted to construct a causative chain of air–sea interaction (Figure 7). The time-lags and the exact energy conversions in this complicated two-fluid system are not yet clear. It is clear, however, that there is an intimate interrelationship between atmospheric and oceanic circulation and energy balance and it seems to be generally accepted now that one cannot study global processes in either of these fluids separately from the other.

We may finally note that just this latter viewpoint of close connection between atmospheric and oceanic energetics has been stressed since decades by the master of oceanography, Georg Wüst, whose 80th anniversary the present volume celebrates. As early as 1920, in his doctoral thesis, he presented quantitative data for the evaporation from the ocean's surface. From there, he proceeded to a detailed picture of precipitation and evaporation over the oceans (Wüst, 1936, 1954) which is still valid and serves as the inherent inte-

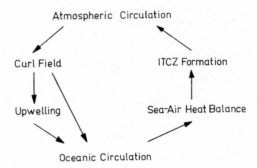

Figure 7   Sketch of sea-air interaction

gral constraint for any discussion of energy processes in oceans and atmosphere on a global scale. The present study may serve as a contribution towards further understanding these highly complex processes and may continue the lines Georg Wüst has initiated.

APPENDIX A

The equations for momentum and mass conservation for a medium on the rotating earth are in the usual notation:

$$\frac{\partial}{\partial t}(\varrho u) + \frac{\partial}{\partial x}(u\varrho u) + \frac{\partial}{\partial y}(v\varrho u) + \frac{\partial}{\partial z}(w\varrho u) - f\varrho v$$
$$= -\frac{\partial p}{\partial x} + A\varrho\,\nabla^2 u + \frac{\partial \tau_x}{\partial z} \tag{A.1}$$

$$\frac{\partial}{\partial t}(\varrho v) + \frac{\partial}{\partial x}(u\varrho v) + \frac{\partial}{\partial y}(v\varrho v) + \frac{\partial}{\partial z}(w\varrho v) + f\varrho u$$
$$= -\frac{\partial p}{\partial y} + A\varrho\,\nabla^2 v + \frac{\partial \tau_y}{\partial z} \tag{A.2}$$

$$\frac{\partial}{\partial t}(\varrho) + \frac{\partial}{\partial x}(\varrho u) + \frac{\partial}{\partial y}(\varrho v) + \frac{\partial}{\partial z}(\varrho w) = 0 \tag{A.3}$$

u, v, w are the velocity components in the x- (positive eastward), y- (positive northward), and z- (positive upward) directions, t is the time. f is the Coriolis parameter, p the pressure, A the eddy viscosity coefficient assumed to be constant, ϱ the density, τ_x and τ_y are the stress components. Eqs. (A.1) and (A.2) are the flux forms of the familiar Navier–Stokes equations. An additional equation for the vertical velocity component w is not needed because this component is eliminated in the following by means of vertical integration.

We now introduce horizontal boundaries in the medium: an upper boundary, the surface $s(x, y, t)$, and a lower boundary, the bottom $b = $ const. which is assumed to be strictly horizontal. Integrating eqs. (A.1) and (A.2) with respect to z from b to s and introducing the abbreviations

$$M_x \equiv \int_b^s \varrho u \, dz \quad M_y \equiv \int_b^s \varrho v \, dz \quad Q \equiv \int_b^s \varrho \, dz \quad P \equiv \int_b^s p \, dz \quad \text{(A.4)}$$

we obtain without neglecting any quantity

$$\frac{\partial M_x}{\partial t} + \frac{\partial}{\partial x} \int_b^s u\varrho u \, dz + \frac{\partial}{\partial y} \int_b^s v\varrho u \, dz - fM_y$$

$$= -\frac{\partial P}{\partial x} + p_s \frac{\partial s}{\partial x} + A \int_b^s \varrho \, \nabla^2 u \, dz + \tau_{xs} - \tau_{xb} \quad \text{(A.5)}$$

$$\frac{\partial M_y}{\partial t} + \frac{\partial}{\partial x} \int_b^s u\varrho v \, dz + \frac{\partial}{\partial y} \int_b^s v\varrho v \, dz + fM_x$$

$$= -\frac{\partial P}{\partial y} + p_s \frac{\partial s}{\partial y} + A \int_b^s \varrho \, \nabla^2 v \, dz + \tau_{ys} - \tau_{yb} \quad \text{(A.6)}$$

$$\frac{\partial Q}{\partial t} + \frac{\partial M_x}{\partial x} + \frac{\partial M_y}{\partial y} = 0 \quad \text{(A.7)}$$

Compare, e.g., Gates (1966), who also takes gradients of b into account. These equations are usually simplified by the following assumptions:

(a) u and v in the nonlinear terms can be replaced by M_x/Q and M_y/Q, respectively, whose functions are independent of z. Hence we set

$$\frac{\partial}{\partial x} \int_b^s u\varrho v \, dz = \frac{\partial}{\partial x} \left(\frac{M_x M_y}{Q} \right) \quad \text{(A.8)}$$

and similar expressions for the three other nonlinear terms. If the velocity components u, v are independent of z in the medium under consideration, eq. (A.8) is exact. If, on the other hand, there exists a temperature gradient in the medium which causes a thermal wind relation, eq. (A.8) neglects the correlation between the u- and v-fields. We further simplify (A.8) by replacing

Q by an appropriate spatial average $Q_0 = \text{const.}$; hence we assume

$$\frac{\partial}{\partial x} \int_b^s u\varrho v \, dz = \frac{1}{Q_0} \frac{\partial}{\partial x} (M_x M_y) \tag{A.9}$$

and similar expressions for the three other terms.

(b) The internal friction terms are parameterized by setting

$$\left. \begin{array}{l} A \displaystyle\int_b^s \varrho \nabla^2 u \, dz = A \nabla^2 M_x \\[3mm] A \displaystyle\int_b^s \varrho \nabla^2 v \, dz = A \nabla^2 M_y \end{array} \right\} \tag{A.10}$$

(c) The bottom stress terms are parameterized by setting

$$\tau_{xb} = kM_x, \quad \tau_{yb} = kM_y \tag{A.11}$$

with k being a uniform constant throughout the fluid.

With eqs. (A.9)–(A.11) we obtain from (A.5) and (A.6):

$$\frac{\partial M_x}{\partial t} + \frac{1}{Q_0} \frac{\partial}{\partial x} (M_x M_x) + \frac{1}{Q_0} \frac{\partial}{\partial y} (M_y M_x) - fM_y$$

$$= -\frac{\partial P}{\partial x} + p_s \frac{\partial s}{\partial x} + A \nabla^2 M_x - kM_x + \tau_{xs} \tag{A.12}$$

$$\frac{\partial M_y}{\partial t} + \frac{1}{Q_0} \frac{\partial}{\partial x} (M_x M_y) + \frac{1}{Q_0} \frac{\partial}{\partial y} (M_y M_y) + fM_x$$

$$= -\frac{\partial P}{\partial y} + p_s \frac{\partial s}{\partial y} + A \nabla^2 M_y - kM_y + \tau_{ys} \tag{A.13}$$

If one deals with this set of primitive equations—namely, eqs. (A.12), (A.13), (A.7)—, one must, in addition to assuming (a) through (c) above, specify the pressure function P. This can conveniently be done by equating

$$\text{grad } P = \frac{g}{\varrho_0} Q \text{ grad } Q \tag{A.14}$$

where ϱ_0 is an appropriate spatial average of ϱ. Eq. (A.14) is easily verified in the case of uniform density.

In order to proceed to the vorticity equation, we assume non-divergence of the motion which allows for introduction of a stream function ψ with

$$M_x = -\frac{\partial \psi}{\partial y} \quad M_y = \frac{\partial \psi}{\partial x} \tag{A.15}$$

Cross-differentiating eqs. (A.11), (A.12) and taking (A.15) into account one obtains in the usual manner, rearranging the several terms:

$$\left(\frac{\partial}{\partial t}\right) \qquad\qquad (J) \qquad (\beta) \quad (A) \qquad\qquad (k) \qquad (\text{curl})$$

$$\frac{\partial}{\partial t}\nabla^2\psi + \frac{1}{Q_0}J(\psi, \nabla^2\psi) + \beta\frac{\partial\psi}{\partial x} - A\nabla^2(\nabla^2\psi) + k\nabla^2\psi = \text{curl}_z\,\vec{\tau}$$

$$\text{(A.16)}$$

J denotes the familiar Jacobian operator, β the derivative of f with respect to y; further,

$$\text{curl}_z\,\vec{\tau} = \frac{\partial\tau_y}{\partial x} - \frac{\partial\tau_x}{\partial y} \tag{A.17}$$

In the derivation of (A.16), it was tacitly assumed that p_s, the air pressure at the sea surface, has no horizontal gradients; this is believed to be a fair approximation for large-scale motions. However, as Gates (1966) indicates, neglection of the pressure torque on the ocean's surface can become crucial on scales of the order of 100 km. The pressure torque as well as the curl term due to bottom topography which we have neglected are both included in Gates' equations as well as in Welander's (1959) generalized transport equation.

The terms in the vorticity equation (A.16) are denoted by short encircled symbols to facilitate discussion in the text.

APPENDIX B

The wind stress data of Hellerman (1967) for the world ocean are the most conscious and most complete compilation of this quantity available. In Hellerman's study the wind stress is computed on the basis of frequency distributions of the surface wind for given geographical locations and time (4 seasons and annual) with the aid of the usual quadratic stress law, the drag coefficient being a function of wind speed and the density a function of latitude. The data are presented as computer printout for most of the $5° \times 5°$-squares over the oceans, in the unit 10^{-2} dynes/cm². For the purposes of the present contribution, the Hellerman data have been transferred to punch cards and processed in a manner fully described elsewhere (Hantel, 1970). A brief description of the procedure applied is given below.

Missing values of the stress components were obtained by linear interpolation in zonal direction, a method that is open to criticism. The completed patterns of τ_x, τ_y were smoothed by an objective filtering method. The filter consists of a discrete weight function of 5 times 5 values; it is symmetrical in both the zonal and meridional directions and is normalized in order to conserve the mean value of the smoothed pattern (Bleck, 1965).

The pertinent wave number response of the filter is a real function and is even in both zonal and meridional directions. Its value is close to unity for wave lengths greater than about five times the grid distance and close to zero for wave lengths less than 2.5 times the grid distance, the grid distance being 5° latitude or longitude. Hence, the filter eliminates all small-scale errors in the original data and retains the important meso- and macro-scale patterns nearly unaltered.

The vertical component of the curl of an arbitrary vector $\vec{\tau} = (\tau_x, \tau_y)$ is

$$\text{curl}_z \, \vec{\tau} = \frac{1}{R \cos \varphi} \left[\frac{\partial \tau_y}{\partial \lambda} - \frac{\partial}{\partial \varphi} (\tau_x \cos \varphi) \right] \tag{B.1}$$

R is the radius of the earth, φ and λ geographical latitude and longitude, respectively. An apropriate discretization formula has been used for evaluating eq. (B.1) in finite difference form (Hantel, 1970). For typical stress values of 1 dyne/cm^2 = 0.1 kg/m/sec^2 typical stress curl values are of the order 5×10^{-9} dynes/cm^3 = 5×10^{-8} kg/m^2/sec^2.

Acknowledgments

The author would like to express his sincere gratitude to Prof. G. Wüst under whose guidance he was introduced into physical oceanography, and to Prof. H. Flohn who suggested the climatological implications of this contribution. Thanks are due to Mr. R. Fabec (National Center for Atmospheric Research, Boulder, Colorado), and to Mrs. B. Wallbrecher for punching and programming assistence. The drawings were prepared by Miss B. Eggemann. The final draft of the manuscript was typed by Mr. H.-G. Jürgensmeier.

The numerical calculations were run on the IBM 7090 computer of the "Gesellschaft für Mathematik und Datenverarbeitung", Bonn, Germany.

References

Bleck, R., Lineare Approximationsmethoden zur Bestimmung ein- und zweidimensionaler numerischer Filter der dynamischen Meteorologie, 86 pp., *Institut für Theoretische Meteorologie der Freien Universität Berlin*, 1965.

Bryan, K., A numerical investigation of a nonlinear model of a wind-driven ocean, *Journal of the Atmospheric Sciences*, Vol. 20, p. 594–606, 1963.

Doberitz, R., Cross spectrum analysis of rainfall and sea temperature at the equatorial Pacific Ocean, *Bonner Meteorologische Abhandlungen*, Heft 8, 53 pp., 1968.

Düing, W., Die Monsunströmungen im Indischen Ozean, *Annalen der Meteorologie (Neue Folge)*, Vol. 4, p. 77–82, 1969.

Ekman, V. W., On the influence of the earth's rotation on ocean-currents, *Arkiv för Matematik, Astronomi och Fysik*, Vol. 2, p. 1–53, 1905.

Fischer, G., On a finite difference scheme for solving the non-linear primitive equations for a barotropic fluid with application to the boundary current problem, *Tellus*, Vol. 17, p. 405–412, 1965.

Flohn, H., Bemerkungen zur Asymmetrie der atmosphärischen Zirkulation, *Annalen der Meteorologie (Neue Folge)*, Vol. 3, p. 76–80, 1967a.

Flohn, H., Dry equatorial zones and asymmetry of the global atmospheric circulation, *Bonner Meteorologische Abhandlungen*, Heft 7, p. 3–8, 1967b.

Gates, W. L., On the dynamical formulation of the large-scale momentum exchange between atmosphere and ocean, *Journal of Marine Research*, Vol. 24, p. 105–112, 1966.

Gates, W. L., A numerical study of transient Rossby waves in a wind-driven homogeneous ocean, *Journal of the Atmospheric Sciences*, Vol. 25, p. 3–22, 1968.

Hantel, M., A numerical investigation of the linear primitive equations applied to the wind-driven circulation of the northwestern Indian Ocean between 16° N–16° S, *Annalen der Meteorologie (Neue Folge)*, Vol. 4, p. 73–76, 1969.

Hantel, M., Monthly charts of surface wind stress curl over the Indian Ocean, *Monthly Weather Review*, 1970, in press.

Hidaka, K., Computation of the wind stresses over the oceans, *Records of Oceanographic Works in Japan*, Vol. 4, p. 77–123, 1958.

Hellerman, S., An updated estimate of the wind stress on the world ocean, *Monthly Weather Review*, Vol. 95, p. 607–626, 1967. With correction notice in *Monthly Weather Review*, Vol. 96, p. 63–74, 1968.

Mazeika, P. A., Mean monthly sea surface temperatures and zonal anomalies of the tropical Atlantic, *Serial Atlas of the Marine Environment*, Folio 16, American Geographical Society, 1968.

Mintz, Y. and G. Dean, The observed mean field of motion of the atmosphere, *Geophysical Research Papers*, No. 17, Geophysics Research Directorate, Cambridge, Mass., 65 pp., 1952.

Moore, D. W., Rossby waves in ocean circulation, *Deep-Sea Research*, Vol. 10, p. 735–747, 1963.

Munk, W. H., On the wind-driven ocean circulation, *Journal of Meteorology*, Vol. 7, p. 79–93, 1950.

Pike, A. C., The inter-tropical convergence zone studied with an interacting atmosphere and ocean model, *Rosenstiel School of Marine and Atmospheric Sciences (Division of Atmospheric Science)*, University of Miami, 25 pp., Contract No. F19628-68-C-0144, Scientific Report No. 2, 1970.

Pedlosky, J., A study of the time dependent ocean circulation, *Journal of the Atmospheric Sciences*, Vol. 22, p. 267–272, 1965.

Reid, J. L., Sea-surface temperature, salinity, and density of the Pacific Ocean in summer and in winter, *Deep-Sea Research*, Supplement to Vol. 16, p. 215–224, 1969.

Schmitz, H. P., Modellrechnungen zu winderzeugten Bewegungen in einem Meer mit Sprungschicht, *Deutsche Hydrographische Zeitschrift*, Vol. 17, p. 201–232, 1964.

Stommel, H., The westward intensification of wind-driven ocean currents, *Transactions, American Geophysical Union*, Vol. 29, p. 202–206, 1948.

Stommel, H., Summary charts of the mean dynamic topography and current field at the surface of the ocean, and related functions of the mean windstress, in: *Studies on Oceanography*, Ed. K. Yoshida, p. 53–58, American Edition, University of Washington Press, 1965.

Sverdrup, H.U., Wind-driven currents in a baroclinic ocean; with application to the equatorial currents of the eastern Pacific, *Proceedings of the National Academy of Sciences*, Vol.33, p. 318–326, 1947.

Veronis, G., An analysis of wind-driven ocean circulation with a limited number of Fourier components, *Journal of the Atmospheric Sciences*, Vol.20, p. 577–593, 1963.

Veronis, G., Wind-driven ocean circulation.—Part 1. Linear theory and perturbation analysis;—Part 2. Numerical solutions of the non-linear problem, *Deep-Sea Research*, Vol.13, p. 17–55, 1966.

Veronis, G. and H.Stommel, The action of variable wind stresses on a stratified ocean, *Journal of Marine Research*, Vol.15, p. 43–75, 1956.

Welander, P., On the vertically integrated mass transport in the oceans, in: *The Atmosphere and Sea in Motion*, Ed. B.Bolin, p. 95–101, New York, 1959.

Wüst, G., Die Verdunstung auf dem Meere, *Veröffentlichungen des Instituts für Meereskunde an der Universität Berlin, Neue Folge, Reihe A*, Heft 6, 95 pp., 1920.

Wüst, G., Florida- und Antillenstrom. Eine hydrodynamische Untersuchung, *Veröffentlichungen des Instituts für Meereskunde an der Universität Berlin, Neue Folge, Reihe A*, Heft 12, 48 pp., 1924.

Wüst, G., Oberflächensalzgehalt, Verdunstung und Niederschlag auf dem Weltmeere (nebst Bemerkungen zum Wasserhaushalt der Erde), in: *Länderkundliche Forschung, Festschrift für Norbert Krebs*, Ed. H.Louis and W.Panzer, p. 346–359, J.Engelhorns Nachf., Stuttgart, 1936.

Wüst, G., Gesetzmäßige Wechselbeziehungen zwischen Ozean und Atmosphäre in der zonalen Verteilung von Oberflächensalzgehalt, Verdunstung und Niederschlag, *Archiv für Meteorologie, Geophysik und Bioklimatologie, Serie A*, Vol.7, p. 305–328, 1954.

Yoshida, K. and Han-Lee Mao, A theory of upwelling of large horizontal extent, *Journal of Marine Research*, Vol.16, p. 40–54, 1957.

Cyclonic Rings formed
by the Gulf Stream 1965–66*

F.C. FUGLISTER

Woods Hole Oceanographic Institution
Woods Hole, Massachusetts, 02543

Abstract A cyclonic ring of Gulf Stream origin is defined to be a member of a special class of eddy; when the meandering Gulf Stream forms a long loop to the right of its downstream direction and the two sides of the loop, with currents flowing in opposite directions, approach each other and come in contact, a ring is formed consisting of a closed segment of the Stream revolving cyclonically around a mass of cold water detached from its former position in the slope water area to the left of the Stream. The first observations of the formation, migration and decay of such rings on seven cruises, covering the period from September, 1965, through February, 1966, are described. The two rings observed started as elliptical shapes, the long axis about 250 km, and gradually changed to circles about 110 km in diameter. The maximum surface currents, on the order of 150 cm/sec, continued strong over the five month period. The rings migrated at a rate of about 10 cm/sec, following irregular, probably anticyclonic paths with a net displacement to the SW. It is estimated that they have a life span of about 12 months. On the bases of a few crude assumptions, using average annual surface current speeds, the monthly net departure of the Gulf Stream from its mean position and a few transport values, it is estimated that from 5 to 8 cyclonic rings form per year.

INTRODUCTION

When the meandering Gulf Stream forms a long loop to the right of its downstream direction, and the two sides of the loop, with currents flowing in opposite directions, approach each other and come in contact, a ring is formed consisting of a closed segment of the Stream revolving cyclonically around a mass of cold water detached from its former position in the slope water area to the left of the Gulf Stream. In order for this to occur, at the

---

* Woods Hole Oceanographic Institution Contribution No. 2489.

137

very least a 500 km length of the Stream, measured along the axis, must be involved in the meander. When first formed, the observed rings have been elliptical not circular in shape. Figure 1 is a schematic drawing showing three stages in the formation of such a ring.

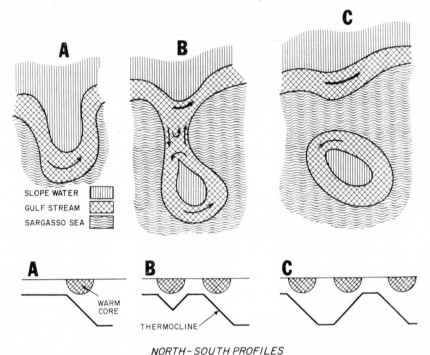

NORTH-SOUTH PROFILES

Figure 1   Schematic of three stages in the formation of a Gulf Stream cyclonic ring

Because the word "eddy" implies some peripheral motion along one side or the other of a boundary or current, it is unfortunate that the term "Gulf Stream eddy" has been used to designate these current rings that are actually made up of long sections including the entire width of the Gulf Stream. Throughout this paper, the word "ring" will be used to differentiate these currents from the smaller scale vortices and eddies associated with shearing motion along the edges of the Gulf Stream (Spilhaus, 1940; Baranov, 1967) and the eddies at current boundaries as illustrated by T. Laevastu (Figure 14, Laevastu, 1962).

In June, 1950, the formation of a Gulf Stream cyclonic ring was observed in a program involving six ships over a period of ten days (Fuglister and Worthington, 1951). That is the only time that this process has been followed in detail, but it seems likely that the many cyclonic "eddies" observed in the past in the northern Sargasso Sea were formed in a similar manner. Most of the evidence for their existence (Iselin, 1936, 1940; Fuglister, 1963; Barrett, 1963) was based on single sections or profiles of temperature measurements made across the anomalously cold water in the Sargasso Sea; only twice in the past have the rings actually been circumnavigated to show their dimensions and to demonstrate that they were separate entities (Iselin and Fuglister, 1948; Fuglister and Worthington, 1951). More recently, Howe and Tait (1967) reported a similar phenomenon in the northeastern Atlantic.

It may be that the short term, small scale implications of the word "eddy" were, in part, why over the years no observational program was specifically designed to study these Gulf Stream rings. Because of the expense and manpower involved, it did not appear feasible to mount a multi-ship operation or a long series of single ship cruises to search for a newly formed ring. The surveys of the Gulf Stream by planes, measuring with infrared thermometry the sea-surface temperature, had not reached the stage where they could be expected to observe the details of ring formation nor did they extend far enough to the east where this phenomenon occurs.

Two events in 1964–65, however, motivated the first study of Gulf Stream rings. First, a new method of tracking the Gulf Stream (Fuglister and Voorhis, 1965) made it possible to map the path of the Stream rapidly over long distances by towing a temperature sensor at a depth of 200 meters. Second, the Environmental Science Services Administration (ESSA) planned to use this method to plot the path of the Gulf Stream, from Cape Hatteras to about 55° west longitude, once a month for a year, starting in August 1965. With ESSA keeping track of the Stream, and with this faster method of detecting and following currents available, it seemed an ideal time to attempt to answer some of the questions posed after the 1950 multiple ship operation. Is there a preferred point of origin for this type of ring formation? How often do the rings occur? Once formed, do they move away from the main current and in what direction? How long do they last? Are elongated rings typical? Do elongated rings break up into smaller, more circular ones? Do these rings ever become re-absorbed by the main current?

We planned to make two consecutive three-week cruises starting in mid-September, 1965. By this time we expected that ESSA would be tracking the Gulf Stream for the second time and that their findings would help us to determine where to pinpoint our search for a ring-forming meander. However, the ESSA program was delayed and we had to start our work without this information. After finding two rings on the first try, it soon became evident that we needed more ship time than we had planned on and the program was extended to the limit of available ship time. We made six cruises on the *RV Crawford* and one on the *RV Atlantis II*, covering the period from the 7th of September, 1965, to February 25, 1966.

This being the first attempt to observe the behavior of cyclonic rings, a brief chronology of the cruises will serve to show some of the problems of this kind of a study and how the picture, as a whole, slowly evolved.

SEQUENCE OF OBSERVATIONS

First cruise, Sept. 8–20, 1965

The method of following the currents was as described by Fuglister and Voorhis (1965). The track of the thermistor, towed at a depth of 200 m, and the positions of all the BT observations (250 m) from the first *Crawford* cruise are shown in Figure 2. The dashed line represents the path of the Gulf Stream (15°C at 200 m) from the U.S.C.&G.S.S. *Explorer*, September 11–18, 1965; this is a smoothed line and does not show the ship's track except for the short segment near 38° N, 61° W.

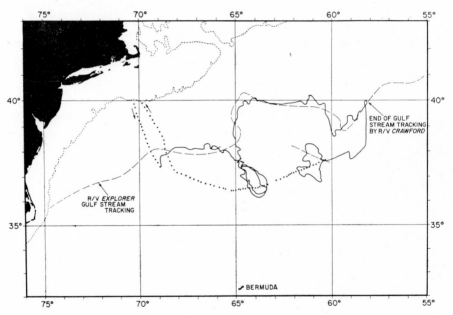

Figure 2   Track of thermistor towed at a depth of 200 m (solid line) and position of BT observations (dots) from *Crawford* cruise 130, Sept. 8 to 20, 1965. R.V. *Explorer* positions (dashed lines) see text

After all our concern with the probable difficulties of finding a cyclonic ring, especially of locating a newly formed one, on the first cruise a meander of the Gulf Stream was observed, near 64° west longitude, that was in the process of forming a cyclonic ring. After following the 200 m temperature gradient (approximately the line of maximum surface current) around the meander, the *Crawford* went around again, following the sharp surface temperature gradient (approximately the left hand edge of the current). On September 12th a short BT section was made across the narrow neck of the meander which showed virtually no surface cold water and only an 18 km wide band of cold water at 200 m. Three days later the *Explorer* tracked the Gulf Stream through this area and saw no indications of a ring-forming meander. There can be no doubt that the Gulf Stream ring that formed near 64° W had broken away from the main current by the 15th of September.

After the 12th of September the *Crawford* continued to follow the Gulf Stream toward the east, going as far as 58° W. Near 60° W, a deflection in the path of the current was observed that appeared to be the remnants of a ring-forming meander; consequently, after leaving the main current, the ship was headed south of this area to see if indeed a ring was in the vicinity. On the 18th a mass of cold water, typical of a newly formed cyclonic ring, was found and circumnavigated. There was clearly a large, irregularly shaped mass of cold water centered approximately at 37° 30′ N, 61° 05′ W. As shown on Figure 2, a BT section was made through this ring (hereafter referred to as the "eastern ring") and then through the "western ring" as the *Crawford* returned to Woods Hole. All the observations along this track, south of the Gulf Stream, except within these rings, showed typical Sargasso Sea temperatures. Certainly, other rings may have been present, the purpose of showing this and the following track charts is not only to show the sequence of our observations, but to show the limits of our coverage. The BT section through the western ring on the 19th showed that this ring, after breaking away from the main current, had not moved east or west in 7 days.

On the *Explorer*, the deflection of the Gulf Stream path near 60° W had also been observed and on the return trip, BT observations were made south of that point; the location where the *Explorer* crossed the eastern ring is indicated by the short dashed line in Figure 2.

At the end of the first cruise it was clear that two cyclonic rings existed south of the Gulf Stream between 59° and 69° west longitudes. The western one was formed—became detached from the Stream—between the 12th and 15th of September; the eastern one was formed earlier, but how much earlier is uncertain. The eastern ring was more extensive, but its actual shape was not clearly determined. Because of the exploratory nature of this first cruise and for lack of time no deep stations were made.

Second Cruise, Sept. 28–Oct. 13, 1965

The positions where observations were made on the second *Crawford* cruise are shown in Figure 3. Except for a short run (71° W to 68° W), no attempt was made to follow the path of the Gulf Stream, but rather the aim was to stay south of it to observe any anomalously cold water that might be in the area. No such water was found until the ship reached near 62° W where the temperature at 200 m dropped below 15 °C. A survey of this cold water area showed an irregular, elliptical ring approximately the same size as the eastern

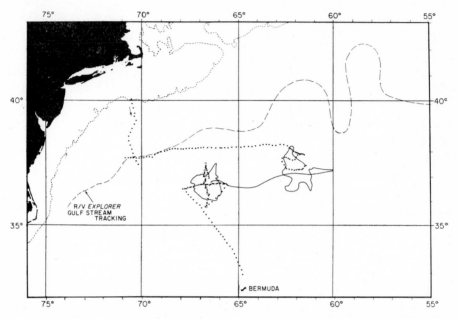

Figure 3 Positions of the observations made on *Crawford* cruise 131, Sept. 28 to Oct. 13, 1965. Position of stations (o). Gulf Stream track by R.V. *Explorer*, Oct. 6–13 (dashed lines)

ring observed on the first cruise. Even as it was being surveyed it was evident that the ring was moving westward. The 200 m temperature field as deduced from this survey is shown in Figure 4; the "center" of the ring had moved 89 km WSW from where it was last observed 16 days earlier. Because of its general size and shape, its similar temperature structure and the direction it was moving between the 2nd and 7th of October, there can be no doubt that this was the same "eastern" ring, and that it had moved toward the WSW on average at a rate of 6.5 cm/sec.

After leaving this ring the *Crawford* passed through the area where the western ring was last observed and on the 8th of October (see Figure 5) entered a cold water mass 230 km to the west. Assuming that this was the western ring, we conclude that it had moved westward at 12.7 cm/sec. After circumnavigation of this cold water mass, a north–south section of eight oceanographic stations and an east–west bathythermograph run were made

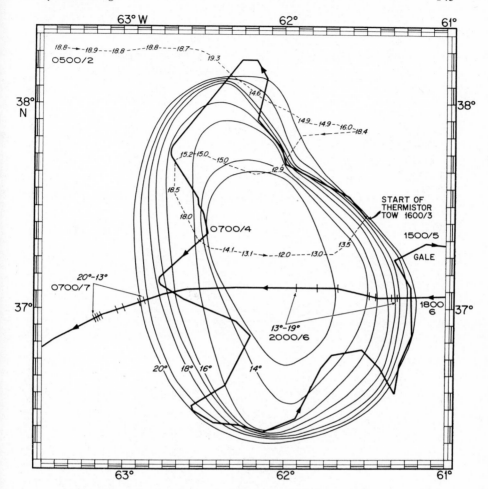

Figure 4 Ships track and 200 m temperatures in the eastern ring, Oct. 2–7, 1965

through the ring. On the western side of the BT run, moderate seas and good Loran conditions made it possible to calculate surface current velocities; the average speed was 124 cm/sec toward the south. The vectors (ranging in magnitude from 62 to 216 cm/sec) are shown in Figure 5.

Third cruise, Oct. 19–30, 1965

At this point in the study, it seemed evident that both rings were moving SW along a path roughly parallel to the Gulf Stream. In order to intercept the western ring the *Crawford* was headed for a point SW of the last known position, see Figure 6. Finding no evidence of a ring by the time we reached 36° N, the ship was headed NE toward where the western ring was last

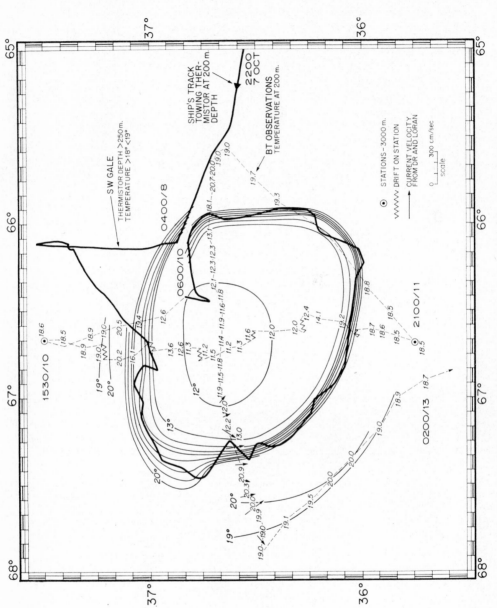

Figure 5   Ships track and 200 m temperatures in the western ring, Oct. 7–13, 1965

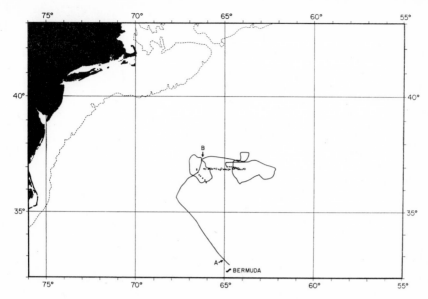

Figure 6 Positions of the observations made on *Crawford* cruise 132, Oct. 19–30, 1965. Continuous temperature from A to B are shown on Figure 7

observed. On the 21st of Oct., it was found 88 km NNE of its former position, having moved at an average rate of 4.0 cm/sec between the 9th and 21st.

An example of the type of information obtained on these runs is shown in Figure 7 which is the record of the run between points *A* and *B* shown on Figure 6. The Sargasso Sea temperatures are relatively uniform, as expected; over a distance of 550 km the temperature at 190 m varied $< \pm 0.8°$, the surface temperature $< \pm 0.5°$. The right side of Figure 7 clearly shows the

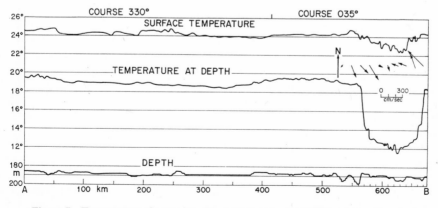

Figure 7 Temperatures from a bow thermistor and a towed thermistor, *Crawford* cruise 132. Ships track A to B shown in Figure 6

abrupt temperature changes that took place when the ship reached and passed through the western ring. The fluctuations in the depth of the towed thermistor when the ship entered the ring are typical. The current vectors shown in the figure are based on only moderately good, hourly Loran fixes; the speeds vary from 0.5 to 289 cm/sec.

Going to the east, the *Crawford* next encountered the cold water of the eastern ring and, with a great deal of difficulty because of high winds, mapped out its position. Compared to the western ring, this was still a larger, less well-defined elliptical shape with the long axis now oriented east–west. It was estimated that the "center" of this ring was continuing to move toward the WSW at a rate of 7.8 cm/sec.

A line of stations was started going from the "center" of the eastern ring toward the "center" of the western one (Figure 6). The section was not completed because a NW gale forced the ship to head back to Bermuda, but enough stations were made to produce some wholly unexpected results. It appeared as though the two rings had joined together, or were in the process of doing so. The eight stations, see Figure 6, averaging 3000 m deep and 29 km apart, interspersed with BT observations, all showed relatively cold water with the 10° isotherm shallower than 500 m.

Fourth Cruise, Nov. 5–21, 1965

Returning north on the 6th of Nov., see Figure 8, the *Crawford* passed through the area where the trough of cold water had been located on the 29th of Oct. The ship reached the southern edge of the Gulf Stream without encountering any anomalously cold water. Turning east, the *Crawford*, late on the 7th, at 64° W long., finally entered cold water. After crossing to the east, the extent of this water was mapped using the towed thermistor. It was an elliptical ring with a long axis oriented north–south.

Still towing the thermistor, the *Crawford* next headed west through typical Sargasso Sea water and at a point 45 km to the west of where the ship had passed heading north 4 days earlier, cold water was again encountered. This proved to be a nearly circular cyclonic ring.

Thus on the 13th of Nov., 15 days after the two cyclonic rings had appeared to join together, there were again two separate rings in the area; the size, shape and thermal structure of both indicated that they were the same western and eastern rings. Under this interpretation, the western ring had moved WSW at a rate of 4.3 cm/sec and the eastern ring NW at 1.3 cm/sec.

A west–east section consisting of 10 stations (3500 m deep) was made through the western ring and then, after a series of GEK measurements, 5 neutrally buoyant floats were placed on a radial line to the west from the "center" at depths of 2000 m. Two days were spent launching and checking the floats, but no meaningful results were obtained before another gale struck. When the ship returned to the area 53 hours later, only two floats were located, and they were no longer in the cold core of the ring. It appeared as though the ring had moved away and that the floats had remained essentially in their original positions. However, there is a distinct possibility that the two floats were incorrectly identified and were in fact two that had moved east during the storm.

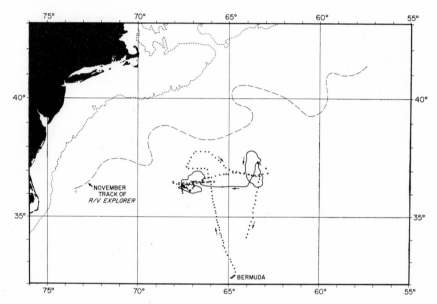

Figure 8 Positions of the observations made on *Crawford* cruise 133, Nov. 5–21, 1965

Before returning to Bermuda, a station (3570 m deep) was occupied at 36° 51′ N, 65° 06′ W, approximately half way between the two rings. Typical Sargasso Sea conditions were observed; the 10° isotherm (the center of the thermocline) was at 850 m, and the 200 m temperature was 19.2°. Then a station (3340 m deep) was made in the eastern ring at 36° 48′ N, 63° 18′ W: the 10° isotherm was at 515 m and the 200 m temperature was 15.1°.

It was evident at this point that two separate cyclonic rings existed in the area. It seems rather fanciful to suppose that these rings would join together to form one ring and then, shortly afterwards, separate again into two, yet this is the simplest interpretation of the admittedly inadequate series of observations of the third and fourth cruises.

Fifth cruise, Nov. 30–Dec. 15, 1965

On the fifth and last cruise in 1965, the *Crawford* headed for the position where the western ring had been last observed (see Figure 9). No cold water was found in that place, but on Dec. 3rd the ring was located approximately 130 km ENE of its Nov. 13th position. The shape was still circular, but it was very much smaller.

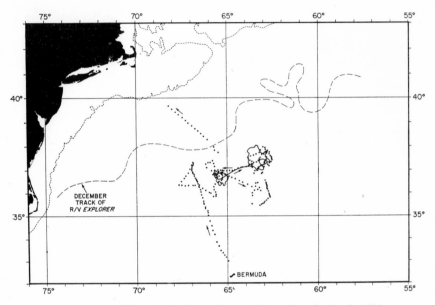

Figure 9   Positions of the observations made on *Crawford* cruise 134,
Nov. 30 – Dec. 15, 1965

Next, a search was made for the eastern ring, but this time not toward its last known position but rather to the south of it. It is apparent that we could not predict the movements of these rings, but these searches at least showed that no other rings existed over a considerable area. Finally on Dec. 7th the eastern ring was located 34 km to the NE of its Nov. 9th position. The ring was mapped and again was found to be roughly elliptical; the long axis oriented NW–SE.

Before the ship headed for Woods Hole, the western and then the eastern ring were again circumnavigated. The western ring was found to be moving toward the SE at 6.1 cm/sec and the eastern ring was continuing toward the NE at 5.2 cm/sec.

Sixth cruise, Jan. 4–23, 1966

After an interval of 19 days instead of the average 6 days, the 6th cruise of the series got underway aboard the *Atlantis II*. On the 5th of Jan. the Gulf Stream (15° at 200 m) was crossed at the position marked *A* in Figure 10, 20 km south of what later developed into the January track of the 15° isotherm by the *Explorer*.

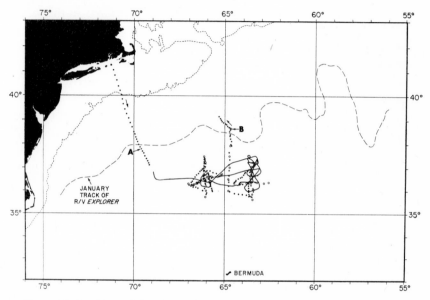

Figure 10   Positions of the observations made on *Atlantis II* cruise 18, Jan. 4–23, 1966

At 36° 30′ N (see Figure 10) the *Atlantis II* was turned to the east to start the search for the rings. A 40 knot southwester slowed progress, but in spite of the generally rough seas it was the increased roughness on the evening of Jan. 6th that was the first indication that the ship was entering the currents of the western ring. Less than an hour after the increased motion of the ship was felt, the temperature at 200 m dropped abruptly below 15° and the Loran fixes showed that the ship was being set to the south. Because of the sea conditions no attempt was made at this time to circumnavigate this ring. It had moved SW at a rate of at least 2.5 cm/sec since last seen on Dec. 12th. At 0600 the following morning, Jan. 7th, the same sequence of events occurred. The sea became noticeably rougher and 50 minutes later the temperature at 200 m dropped below 15° and the ship was in the eastern ring. Less than 3 days after leaving port, both rings had been located. The change in sea state when the ship entered the currents of these rings had occasionally been noticed on board the *Crawford* but never so dramatically as on these two occasions on the *Atlantis II*. The seas were always rougher; that is, more white caps were seen in the portion of the rings where the currents opposed the winds.

With some difficulty because of increasing winds (over 50 knots), the eastern ring was mapped and found to be no longer markedly elliptical but more nearly circular. The "center" had moved at a rate of at least 3.1 cm/sec toward the SSW since it had been last observed in December. At this stage, with the gradual sinking of the thermocline in the center of the ring, it was necessary to make observations a little deeper than 200 m in order to map out the position of the ring, and frequent BT observations to 250 m were made to supplement the continuous record from the towed thermistor.

The wind increased to over 60 knots as the *Atlantis II* returned to the western ring, but on the 10th it moderated enough so that station work could start. Stations about 30 km apart, with sampling to the bottom, were made from north to south through the center of the ring. The "center" of the ring had moved WNW at a rate of 6.6 cm/sec over the past 5 days. Next, an east–west line of stations was occupied and showed an even faster rate of trans-lation of the ring to the WNW, 11.3 cm/sec.

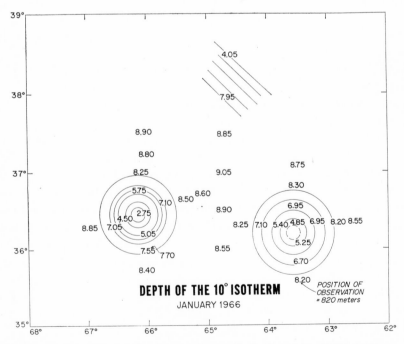

Figure 11 Positions of the deep stations and depths of the 10°C isotherm, Jan. 10–23, 1966

Returning to the east, the *Atlantis II* passed through the area where the eastern ring was located 8 days earlier. There was no current indication at all in the area so, assuming that the ring was continuing on its former path toward the SSW, the ship was turned toward the SW. For the third time on this cruise, the first indication that the ring had been found was the added roughness of the sea; a half hour after this was noticed, the 200 m temperature that had been steady at 18.5 °C suddenly dropped to below 15 °C. This eastern ring had continued to move SSW at a rate of 6.8 cm/sec. The program of observations that had been carried out in the western ring was now repeated in the eastern ring.

Before returning to Woods Hole a line of deep stations, spaced 55 km apart, was run from south to north midway between the two rings. This profile showed typical Sargasso Sea structure, with the northern end in the Gulf Stream (the position marked B in Figure 10); there was no evidence of any connection between the two rings. The locations of all the stations made on this cruise are shown in Figure 11, which also shows the depth of the 10° isotherm at each station.

Seventh cruise, Feb. 4–24, 1966

Back again on *Crawford*, the last cruise of the series started with a BT section (250 m) from the continental shelf SSE to the Gulf Stream (Figure 12). In the Stream the towed thermistor was launched and throughout the remainder of this cruise, with a brief stop in Bermuda, the recording from this instrument together with those of the bow thermistor and frequent Loran fixes were the only observations made. Because of the generally very rough sea state, progress was slow and the towed thermistor was usually at about 300 m depth.

The Stream was followed from the position marked A to B (Figure 12): then the ship headed south to pass through the last known position of the eastern ring. Unfortunately, the ship's track was slightly to the east of that position and no cold water was observed. After much delay because of head winds and considerable trouble with uncertain Loran fixes, the ship was headed west and then north and on the 9th entered the cold water of the ring. It was evident, at this point, that the ship had passed through the east side of the ring on its way south, but the uncertain navigational evidence of this and been disregarded.

The cyclonic currents of the ring were still quite strong, averaging a little over 100 cm/sec. The "center" of the ring had moved at a rate of at least 3.9 cm/sec over a period of 19 days.

The winds moderated as the *Crawford* next went in search of the western ring. With the shrinking size of the rings, the search pattern had to be a close one. After going as far west as 67°30′ W with no signs of a ring the ship made a return track 40 km to the south and on the 11th entered the cold water of the western ring. Its "center" had moved SE at a rate of at least 3.7 cm/sec over a period of 29 days.

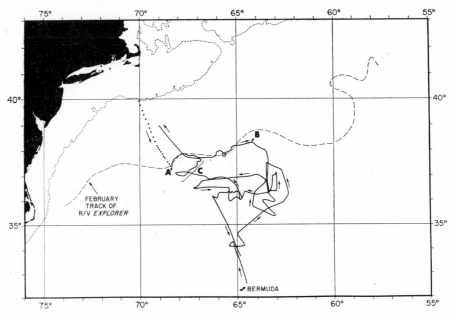

Figure 12   Positions of the observations made on *Crawford* cruise 136, Feb. 2–24, 1966

After mapping this ring, an attempt was made to survey the area to the west, but strong adverse winds prevented it and the ship was headed for Bermuda. On the way to Bermuda near 34° N, for the first time in this series of cruises, anomalously cool water was observed that was not associated with the two rings. The temperature decrease in this "fossil" ring was not very marked; slowing the ship so that the thermistor sank to 400 m gave a minimum temperature of 15.2 °C. The Loran navigation showed no evidence of surface currents. Because of the weather, no attempt was made to map the area.

After a two-day stop in Bermuda, the *Crawford* returned to 34° N, 65° W and again the anomalously cool water was observed. A brief survey of the area showed no sharp horizontal temperature gradients in the surface layer (300 m), nor were there any marked surface currents.

The original intent of this last cruise of the series was to cover a wide area to the east and west to see if any new Gulf Stream rings were in evidence;

but because of continuing foul weather, poor Loran fixes in the eastern sector, and time limitations, this plan was abandoned, and after reaching 62° 15′ W the ship was turned back to obtain two last fixes on the known rings.

Going north, west and then south through the last position of the "center" of the eastern ring, it was again navigational evidence, though questionable, that indicated where the ring was. The ship had gone around the ring; turning east and then north between the two tracks, the cold core was found on the 19th. The "center" of the ring had moved NE at a rate of at least 6.5 cm/sec.

On the 20th the wind moderated and the *Crawford* moved west to obtain a last fix on the western ring. Early in the morning of the 21st the ship entered the cold core of the ring and at the same time the wind increased to about 40 knots from the NW. The ship could not be maneuvred to survey the ring, but simply jogged into the wind, slowly crossing the cold core, a distance of about 44 km. Assuming that the ship crossed the "center" of the ring, it had moved NE at a rate of 5.0 cm/sec. Late on the 22nd, near 37° N, 68° W, cold water was observed and it was assumed that the ship was in the southern part of a new cyclonic ring. Coming about and heading east in order to circumnavigate the ring it soon became evident that the ship was actually in the Gulf Stream, and after following it for approximately 120 km (the segment marked C in Figure 12), the *Crawford* crossed into the slope water and then to Woods Hole to end the 1965–66 series of cruises.

CONCLUSIONS

Movement of the rings

For the following reasons, we conclude that the two rings observed during this program were always the same western and eastern rings:
a   The relative size and shape of the two rings; for the first three months the western one was smaller and more nearly circular.
b   The continuously decreasing size of both rings, as measured at a depth of 200 m.
c   The thermal structure in the central portion of the rings; this changed gradually with none of the fluctuations that would be expected if new and different rings were being observed.
d   The speed of translation of the rings (see Table 1); on the few occasions when the speed was measured over a period of less than a week, it averaged beetwen 7 and 9 cm/sec. Although the evidence for the continuous curvature to the right is slight, plotting such a smooth curve (see Figure 13) results in

Table 1   Translation rates of the ring *centers* (cm/sec)

Lat. N	Long.W	Time	Straight line	Curved line
The Western Ring				
36°45′	64°05′	0900/19/IX		
36°30′	66°35′	2000/9/X	12.8	13.3
36°50′	66°20′	2200/21/X	4.2	7.8
36°39′	66°25′	1100/29/X	3.4	3.6
36°30′	67°00′	1200/13/XI	4.3	5.6
36°55′	65°40′	1000/3/XII	7.4	8.3
36°40′	65°15′	0200/12/XII	6.1	6.7
36°20′	65°45′	1600/6/I	2.5	3.5
36°26′	66°03′	2000/11/I	6.6	6.6
36°30′	66°13′	1300/13/I	11.3	11.3
36°10′	65°15′	1500/11/II	3.7	8.6
36°25′	64°56′	2400/20/II	5.0	6.4
			average 6.1	7.4
The Eastern Ring				
37°30′	61°05′	1000/18/IX		
37°10′	62°00′	1000/4/X	6.5	6.5
37°05′	62°10′	0200/7/X	7.6	7.6
36°55′	63°30′	0500/25/X	7.6	8.3
37°00′	63°40′	1800/9/XI	1.3	11.3
37°25′	63°10′	1200/9/XII	2.4	9.0
37°30′	63°20′	1000/13/XII	5.2	5.4
36°53′	63°28′	1500/8/I	3.1	9.5
36°20′	63°32′	2300/18/I	6.8	12.8
36°12′	63°40′	1200/21/I	8.4	8.4
36°47′	63°40′	1200/9/II	3.9	6.9
37°00′	63°05′	2000/19/II	6.5	6.6
			average 5.4	8.4

average translation speeds of 7.4 cm/sec for the western ring and 8.4 cm/sec for the eastern one.

e   If cyclonic rings generally translate at speeds comparable to these, and we consider the distribution in time and space of all the observations both in the rings and in the surrounding area, it is difficult to see how we could have been mistaken in the identity of the two rings observed.

Another way of depicting the movement of the rings, and also their relationship to the meandering Gulf Stream is shown in Figure 14. The solid line is the position of the 15 °C isotherm at a depth of 200 m, the shaded areas

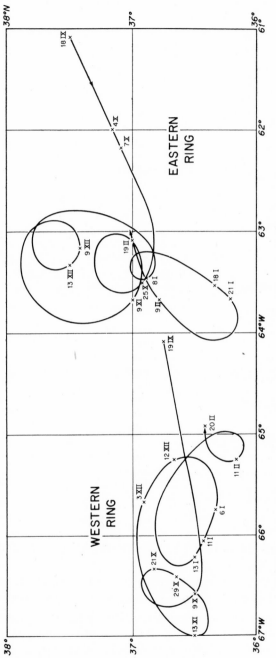

Figure 13 Trajectories of the "centers" of the two cyclonic rings from Sept. 18, 1965 to Feb. 20, 1966

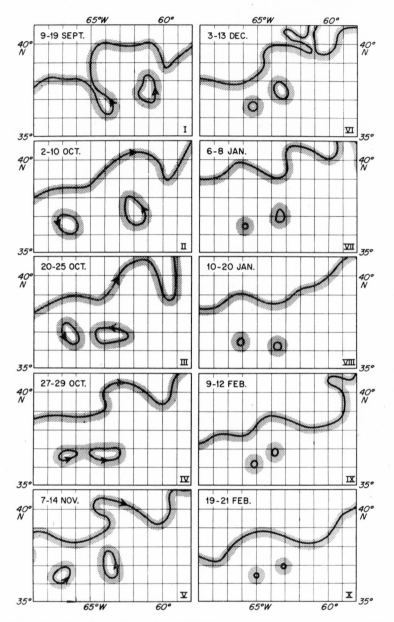

Figure 14 Varying positions of the Gulf Stream and the two cyclonic rings. The heavy line is the approximate position of the 15 °C isotherm at 200 m, the shaded area the estimated minimal width of the currents

represent the estimated minimal width of the currents; facing downstream, the 15°C line is about 20 km from the left hand edge and 60 km from the right hand edge. The method used to determine the 15°C position in the rings was illustrated in Figures 4 and 5. With the rings continuously shifting their positions and the amount of interpolation of the data required to draw this 15°C line, it is obvious that we can only approximate the size and shape of these rings. The paths of the Gulf Stream are mainly based on the *Explorer* data with some interpolation to fit in various points obtained by our vessels.

The charts in Figure 14 clearly show the changing shapes and decreasing sizes of the two rings. There does not appear to be any correlation between the changing path of the Gulf Stream and the movement of the rings. During the last four months the rings were relatively close together and showed no signs of coalescing; the depth of the main thermocline (see Figure 11) clearly shows two separate rings. The apparent joining of the two rings in late October (chart IV, Figure 14) is questionable but it is the only simple interpretation of the data obtained at the end of the third cruise.

Although the two rings, at first, moved in a direction contrary to the Gulf Stream flow, after 5 months there is no indication of their moving out of the general area in which they were formed.

Rate of decay of the rings

The first and last bathythermograph sections made through the two rings are shown in Figure 15. In this four month interval the surface water cooled approximately 6°C; the last survey, a month later when no BT observations were made, showed a further decrease of 1°C in the surface temperatures. The seasonal cooling extended down to approximately 200 m in the outer portion of the rings but in the center this cooling reached to only 50 m; in four months the temperature at 100 m increased 5°C. It is apparent that these rings could not have been followed much longer by this method of observing the temperature structure.

A more complete picture of the changing structure of the rings is shown in deep temperature and salinity sections (Figures 16, 17, 18). Here we see that the changes are not simply due to seasonal cooling and vertical mixing. Considerable horizontal mixing must have occurred to cause the changes in salinity in the upper 100 m layer. Over the outer portion of the western ring the salinity at the surface increased $0.2°/_{00}$ in this three month period; in the center it increased $0.8°/_{00}$ both at the surface where the water had cooled 4°C and at 100 m where it had warmed 4°C. These January sections show that although the evidence for the rings was fast disappearing in the upper 200 m layer, the relatively shallow position of the main thermocline and halocline in the center of the rings was still very much in evidence. A linear extra-

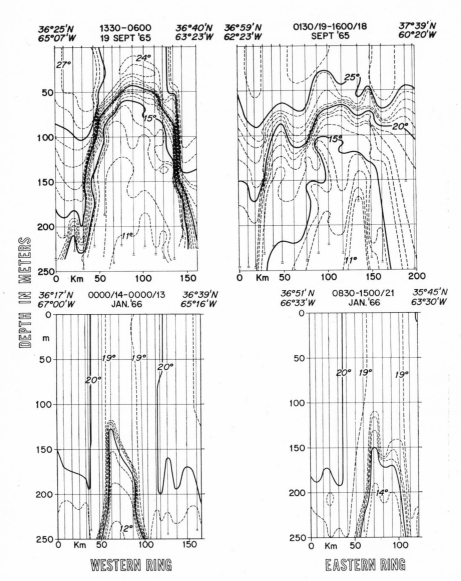

Figure 15   The first and last bathythermograph sections through the western and
eastern rings

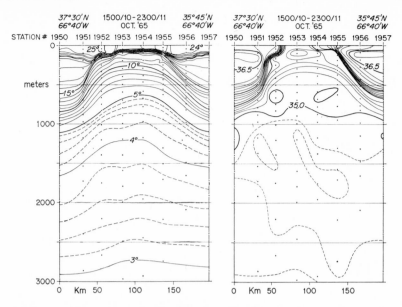

Figure 16 North to South temperature and salinity profiles of the western ring
in October 1965

polation of the rate at which the 10° isotherm was sinking (approximately
1 m/day) in the center of the western ring gives a total life time of 18 months
for this ring. However, it is not the increased depth of the thermocline in the
center but the decreased lateral extent of the ring that is the most obvious
change that occurred. In three months the diameter (10° at 800 m) decreased
approximately 40 km; if the ring continued to shrink at this rate it would
have a total life time of 12 months. Although it may be impossible to determine, by observations, the end point of one of these rings, a life time of the
order of one year is indicated by the observations made to date. This time
could be much shorter if a ring were to be "re-absorbed" by the Gulf Stream,
but this phenomenon has not been observed.

Current velocities

Surface current velocities were measured by steering a fixed course at constant speed across the rings and, with half hourly Loran A positions, finding
the displacements due to the motion of the water. This method always
clearly indicated the cyclonic movement in these rings, but only on a few
occasions were conditions such that reasonably accurate current velocities
could be measured. The maximum current speeds obtained in this manner
were always greater than 100 cm/sec, once as high as 289 cm/sec, and

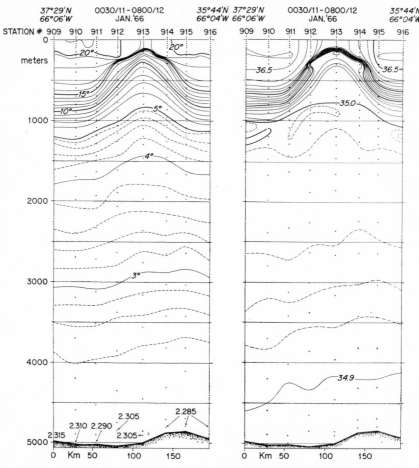

Figure 17 North to South temperature and salinity profiles of the western ring in January 1966

averaged 124 cm/sec. There was no evidence of the currents slowing down over the five months.

One set of GEK measurements, made on the fourth cruise, gave much lower velocities: maxima of only 66 cm/sec. The current speeds computed from the station data, obtained before the section of GEK observations was made, gave a maximum of 120 cm/sec.

The geostrophic velocities were computed in the usual manner from the station data, and then, because of the relatively small radius of curvature in these rings, a correction was applied using the gradient wind formula:

$$\frac{v^2}{R} + fv - fv_g = 0$$

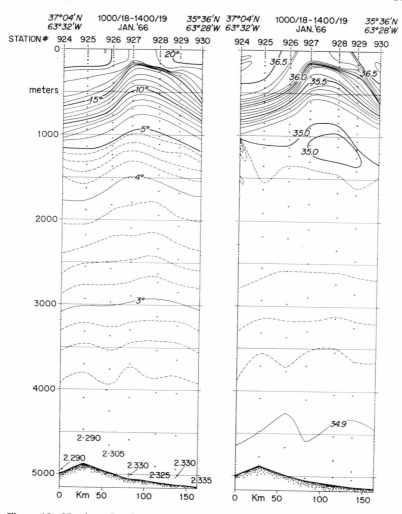

Figure 18 North to South temperature and salinity profiles of the eastern ring in January 1966

where v is the corrected·velocity, R the radius of curvature (positive for cyclonic curvature), f the coriolis parameter and v_g the geostrophic velocity. This had the effect of reducing the velocities by as much as a third in some cases.

Five of the velocity profiles across the rings are shown in Figures 19, 20, and 21. The isotachs are drawn through the computed values plotted midway between station positions. The reference level of 2600 m was used because all of the station data reached at least to this depth. Except at the northern end of the January section through the western ring (Figure 20), the outer limits of the rings are not defined by these profiles.

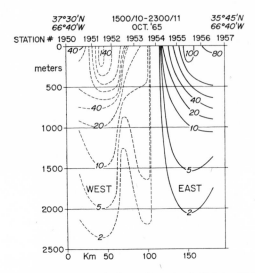

Figure 19   Velocity components in the western ring in October 1965 (cm/sec relative to zero at 2600 m)

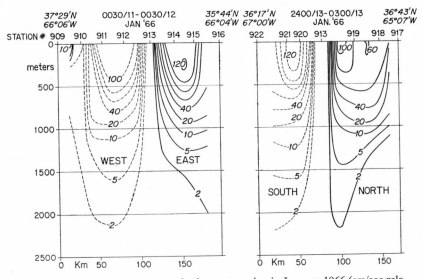

Figure 20   Velocity components in the western ring in January 1966 (cm/sec relative to zero at 2600 m)

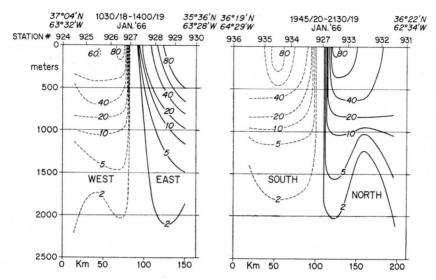

Figure 21   Velocity components in the eastern ring in January 1966 (cm/sec relative to zero at 2600 m)

On the October section (Figure 19) the drift of the ship while on station, as shown in Figure 5, indicated that the northern and southern stations, 1950 and 1957, were out of the current. Because the velocities computed are averages between stations, this is not evident in the velocity profile. Also, the drift of the ship while on station 1954 indicated that this station was south of the center of rotation of the ring and not at the center as shown on the velocity profile. Similar problems arise with using a particular station for the center of both the north–south and east–west profiles shown in Figures 20 and 21. With the constant translation of these rings and their speed of rotation, it is very difficult to plan a series of deep stations (necessarily time-consuming) so as to get a reliable three-dimensional picture. These profiles in general confirm the surface velocities obtained by dead reckoning and Loran. They indicate that the current speeds did not diminish markedly over a three month period, and they show that the currents in the eastern ring were weaker than those in the western ring.

Two deep current meter records were obtained on the January cruise. One was a 60 hour record, 1900/12 to 0700/15, at 36°56′ N, 66°08′ W near station 911. At a depth of 4920 m it showed an average speed of 4.8 cm/sec, the direction varying between 318° and 335°. Even though this speed is less than the estimated translation speed (11.3 cm/sec) of the ring "center" for this period, it is still possible that the direction of this current shows the depth of cyclonic motion in the western ring (see Table 1 and Figures 13, 17).

This is a tenuous assumption however, especially when we consider the current meter record obtained in a similar position relative to the eastern ring. This second meter at 36° 54′ N, 63° 31′ W, near station 925 was set at 1300/18 and picked up at 0800/21. At a depth of 4900 m it showed an average speed of 5.1 cm/sec toward 104°. Here again it is less than the translation speed (8.4 cm/sec) but in this case the direction cannot be associated with a cyclonic motion in the ring, indeed it suggests just the opposite. In the western ring the "center" was moving closer to the current meter during the period of record, in the eastern ring it was moving away but this hardly explains the observed current directions.

These records, together with the general flat appearance of the isotherms and isohalines in the deep waters (Figures 17, 18) indicate that, when these rings were 4 months old, they did not extend down to more than about 3000 m. Unfortunately, 2 deep current meters placed in the southern part of the rings were never recovered.

Frequency of ring formation

Gulf Stream rings not only transfer large volumes of water across an oceanic boundary, thereby mixing different water masses, but they also remove from the Stream itself equally large volumes of water and, in the process of formation, reduce the volume transport of the Stream through a ring generating region. Although the important question of the rate of formation of rings, cyclonic and anticyclonic, by the Gulf Stream cannot be answered on the basis of the data obtained during this 1965–66 study, an order of magnitude estimate will be made.

The path of the Gulf Stream was mapped by ESSA, once a month, over a considerable distance, but aside from the two cyclonic rings that have been dealt with here no positive evidence of ring formation was obtained. However, the paths of the Gulf Stream obtained by ESSA (Niiler and Robinson, 1967; Hansen, in press) showed at least a dozen pronounced meanders that could possibly have formed into rings in the intervals between cruises.

In order to estimate the number of rings formed by the Gulf Stream in a year two simple models will be used; one based on the difference in the average annual inflow and outflow of the Stream in the area between Cape Hatteras and the Grand Banks (75° to 50° west long.) and the other based on the changing area of the slope water in this same region, measured by the monthly net departure of the Stream from its mean position. In both models the Gulf Stream is treated simply as a nonstretching, continuously moving line marking the boundary between the slope water and the Sargasso Sea.

In the first model we state that the Gulf Stream, as represented by this moving line, is entering the region at 75° W at a faster rate than it is leaving it at 50° W with the result that the length of the boundary line in the region is continuously increasing in the form of meanders north and south of a mean position. We now assume that on the average a steady state is achieved by removing segments of the boundary line in the form of cyclonic and anti-cyclonic rings. The number of rings required to do this depends on their size and difference between inflow and outflow speeds. If it takes a 500 km segment of the Stream to form a ring and the difference in speed is 25 cm/sec, in a year the formation of 15 rings would maintain a balance. This average speed difference was obtained from the Current Atlas of the North Atlantic (U.S. Navy Hydrographic Office, 1946). This is a very crude model, especially when we consider that the decrease in speed may be due to an increase in current width or a slow geostrophic outflow from both sides of the Stream, but it is interesting that essentially the same results are obtained in the second model where the current speeds are not considered.

In the second model we assume that the area covered by the slope water, between 75° and 50° west long., is decreased when a cyclonic ring is formed and roughly 18,000 km^2 of slope water is transferred to the Sargasso Sea; conversely an anticyclonic ring increases the size of the area by the same amount. We further assume that, in order to maintain a balance, over a period of a year an equal number of the two types of rings occur.

Changes in the size of the slope water area can be calculated by measuring the north–south displacement of the mean position of the Stream between 75° and 50° W. The mean position of the Gulf Stream for the year (Sept. 1965 through Aug. 1966) was calculated by using the ESSA data on the positions of the 15° isotherm at 200 m. The monthly net departure of the Stream from this mean position for the region between 62° and 71° W long. and for the whole region of interest between 50° and 75° W long. is shown in Figure 22. For this first region, data were available for each degree of longitude, for all twelve months. For the second region, varying amounts of extrapolation were required to carry the Stream path out to 40° N, 50° W. The departure from this same mean position of two paths of the Gulf Stream that covered the whole area, one in June 1950 the other in June 1964, are also indicated.

The distance along the mean path of the Stream from 50° to 75° W long. is approximately 2300 km. The Gulf Stream is displaced 60 km during a year so that the slope water area is a minimum in November and a maximum in April, the difference being 138,000 km^2. If, as we have assumed, this change in area is due to the formation of rings then approximately 8 cyclonic rings are formed between April and November and 8 anticyclonic rings in the remainder of the year.

Iselin (1940) suggested that the Gulf Stream occupies a northerly position when it is weak and a southerly position when it is strong; on the basis of

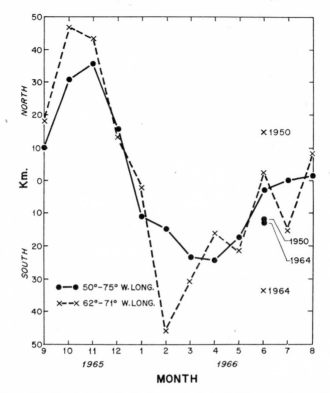

Figure 22 North to South displacement of the Gulf Stream during 1965–66

transport calculations and tide gauge records he showed that the current is relatively strong during the early summer, falls off rapidly in strength to a minimum in October or November and then increases rapidly until January or later. Iselin's data called for a secondary minimum (northerly shift) in April or May, but otherwise his curves suggest the same seasonal trends o the north–south displacement of the Stream as shown in Figure 22. This is remarkable since Iselin had no data on the monthly net departure of the Gulf Stream from its mean position; and there are no data from the ESSA survey to show whether or not there was any seasonal fluctuation in current strength.

The estimate of 16 rings forming per year is based on average temperatures and currents in the surface layer. There exists relatively little data for the deeper water; but, if this water is taken into account, the number 16 appears to be too high. The development of large meanders must reduce the volume transport of the Gulf Stream, but unfortunately all the recent estimates of the volume transport deal with that portion of the system to the west of the 65° W long., where the transport is increasing downstream (Knauss, 1969).

However, some earlier estimates (Mann, 1967; Fuglister, 1963) that are restricted to the transport of the upper 2000 m layer, cover some of the region of decreasing transport. Between the area of maximum transport (65°–70° W) and the tail of the Grand Banks (50° W) various calculations show a decrease of between 30 and 40 million cubic meters per second. Assuming a typical ring is made up of a segment of the Gulf Stream 500 km long, 100 km wide and 2 km deep, it has a volume of 100,000 km^3. If all of the decrease in transport is attributed to the formation of such rings, then 10 to 13 rings are formed per year.

It is quite possible that anticyclonic rings, none of which have been thoroughly surveyed, may, because they move into shallower depths, remove less water from the stream than do the cyclonic rings. If this is so, it would increase the number of rings, coming closer to the earlier estimate of 16.

To conclude these speculations on frequency of ring formations it is estimated that between 5 and 8 cyclonic rings and an equal number of anti-cyclonic rings are formed each year.

Station Data

All of the station data and copies of the bathythermograms from *Crawford* cruises 130–134 and 136, and *Atlantis II* cruise 18 are on file at the National Oceanographic Data Center, Washington, D.C.

Acknowledgements

Except during the first cruise of this program it was a very rough time of year, especially for the relatively small ship *Crawford*, and her Captain, David F. Casiles, her officers and crew are to be commended for excellent work done under very trying conditions. The scientific program was under the direction of the author with Dr. A. D. Voorhis and Mr. C. E. Parker, ably assisted at various stages by a number of oceanographers including Mr. W. G. Metcalf, Mr. L. V. Worthington, Mr. J. R. Barrett, Jr., Mr. G. H. Volkmann, Mr. J. G. Bruce, Dr. H. M. Stommel and Dr. B. A. Warren.

This work was done under the Office of Naval Research, Contract Nonr-2196 (00) NR 083-004 and N00014-66-C-0241; NR 083-004.

References

Baranov, Ye. I., Study of eddies in the Frontal Zones in the Gulf Stream. *Oceanology*, 7 (1), 61–65, 1967.

Fuglister, F. C. and L. V. Worthington, Some results of a multiple ship survey of the Gulf Stream. *Tellus*, 3 (1), 1–14, 1951.

Fuglister, F. C., Gulf Stream '60. *Progress in Oceanography*, I, 263–373, 1963.

Fuglister, F. C. and A. D. Voorhis, A new method of tracking the Gulf Stream. *Limnol. Oceanog.*, Suppl. to Vol. 10, R115–R124, 1965.

Hansen, Donald V., Gulf Stream meanders between Cape Hatteras and the Grand Banks. *Deep-Sea Res.*, in press.

Howe, M. R. and R. I. Tait, A subsurface cold-core eddy. *Deep-Sea Res.*, **14** (3), 373–378, 1967.

Iselin, C. O'D., A study of the circulation of the western North Atlantic. *Pap. Phys. Oceanog. Meteor.*, **4** (4), 1–101, 1936.

Iselin, C. O'D., Preliminary report on long-period variations in the transport of the Gulf Stream System. *Pap. Phys. Oceanog. Meteor.*, **8** (1), 1–40, 1940.

Iselin, C. O'D. and F. C. Fuglister, Some recent developments in the study of the Gulf Stream. *J. Mar. Res.*, **7** (3), 317–329, 1948.

Knauss, John A., A note on the transport of the Gulf Stream. *Deep-Sea Res.*, Suppl. to Vol. 16, 117–123, 1969.

Laevastu, T., The components of surface currents in the sea and their forecasts. *Proceedings of the Symposium on Math.-Hydro. Methods of Phys. Oceanog.* Institut für Meereskunde, Universität Hamburg, 321–338, 2 Tables, 14 Figs, 1962.

Mann, C. R., The termination of the Gulf Stream and the beginning of the North Atlantic Current. *Deep-Sea Res.*, **14** (3), 337–359, 1967.

Niiler, P. P. and A. R. Robinson, Theory of free inertial jets, II. A numerical experiment for the path of the Gulf Stream. *Tellus*, **19** (4), 601–619, 1967.

Spilhaus, A. F., A detailed study of the surface layers of the ocean in the neighborhood of the Gulf Stream with the aid of rapid measuring hydrographic instruments. *J. Mar. Res.*, **3** (1), 51–75, 1940.

Stommel, Henry, *The Gulf Stream*. Univ. of California Press, Berkley and Los Angeles, 1965.

U.S. Navy Hydrographic Office, Current Atlas of the North Atlantic. H.O. Misc., **10**, 688, 1946.

Anticyclogenesis in the Oceans as a Result of Outbreaks of Continental Polar Air*

L.V. WORTHINGTON

Woods Hole Oceanographic Institution
Woods Hole, Massachusetts, 02543

Abstract In winter, continental polar air frequently breaks out over the Pacific Ocean south of Japan and over the Atlantic Ocean south of New England and east of Newfoundland. As a result, in these ocean areas deep isothermal layers are formed in late winter by convergence and sinking of surface water. The greatest thermocline depths and associated high pressure zones are found directly below these convergences. It is suggested that these cold outbreaks are in large part responsible for the general circulation in these regions.

The Northern Hemisphere contains two large land masses, Asia and North America, from which continental polar air flows over the oceans in winter. Outbreaks of this cold, dry air-mass take place at relatively low latitudes. The average Northern Hemisphere surface pressure map for January (Figure 1) gives an indication of where these outbreaks are most likely to take place. Over the northwestern Pacific the major outbreaks occur across the Japan Sea and the islands of Japan; the Aleutian low, which is well developed in winter, inhibits the flow of continental polar air at higher latitudes. Over the northwestern Atlantic the Icelandic low similarly restricts the outbreaks to a region between Newfoundland and the Carolinas.

These outbreaks result in extremely high heat transfer from the oceans to the atmosphere. The mean evaporative heat transfer for January, according to Budyko (1963) is shown in Figure 2; the units are 10^3 cal/cm^3/mo. In the Pacific a large region of high heat flux appears south of Japan; in the Atlantic two smaller regions appear south of New England and east of Newfoundland. The effect of this on the oceans is that deep isothermal layers are

---

* Woods Hole Oceanographic Institution Contribution No. 2499.

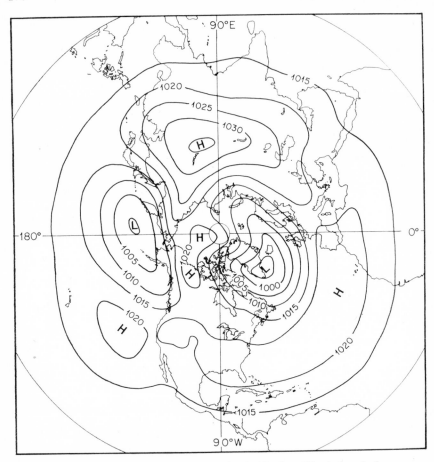

Figure 1   Normal sea-level pressure (mb.) for the Northern Hemisphere in January.
After U.S. Weather Bureau Technical Paper No. 21 (1952)

formed in these areas in late winter. Temperature/depth curves for these
regions in both winter and summer are plotted in Figure 3. In the Pacific
curve 1 represents winter conditions south of the Kuroshio; there is an isothermal layer at least 308 m deep at a temperature of 17.8 °C. Curve 2 shows
summer conditions; a seasonal thermocline has developed but the distinct
thermostad at 18 °C persists through the summer above the main thermocline.
Curve 3 is from the eastern Pacific and is included to show the temperature
conditions which prevail in areas where continental polar air does not reach.
A similar state of affairs exists in the Atlantic south of the Gulf Stream.
Curve 4 (late winter) shows a somewhat deeper isothermal layer than is
found south of the Kuroshio and curve 5 (late spring) shows the early
development of the seasonal thermocline with the same characteristic

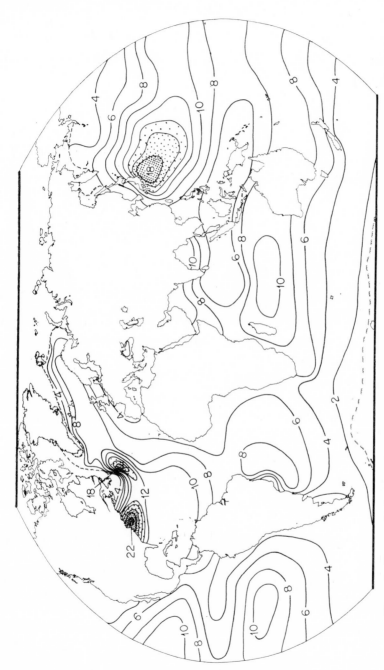

Figure 2 Evaporative heat flux from the ocean to the atmosphere (10^3 cal/cm^2/mo.) for January. After *Budyko* (1963)

thermostad at 18 °C between the seasonal and main thermocline. Curve 6 is from a late winter station in the eastern Atlantic beyond the reach of continental polar air and the isothermal layer does not exceed 100 m. East of Newfoundland curve 7 shows very late winter conditions; the surface

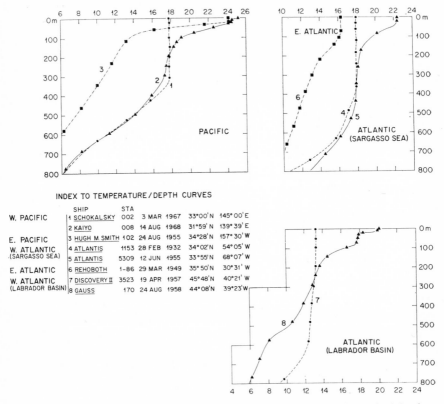

INDEX TO TEMPERATURE / DEPTH CURVES

		SHIP		STA			
W. PACIFIC	1	SCHOKALSKY	002		3 MAR 1967	33°00'N	145°00'E
	2	KAIYO	008		14 AUG 1968	31°59'N	139°39'E
E. PACIFIC	3	HUGH M. SMITH	102		24 AUG 1955	34°28'N	157°30'W
W. ATLANTIC (SARGASSO SEA)	4	ATLANTIS	1153		28 FEB 1932	34°02'N	54°05'W
	5	ATLANTIS	5309		12 JUN 1955	33°55'N	68°07'W
E. ATLANTIC	6	REHOBOTH	1-86		29 MAR 1949	35°50'N	30°31'W
W. ATLANTIC (LABRADOR BASIN)	7	DISCOVERY II	3523		19 APR 1957	45°48'N	40°21'W
	8	GAUSS	170		24 AUG 1958	44°08'N	39°23'W

Figure 3 Temperature/depth curves for the North Pacific and North Atlantic Oceans

layer is not truly isothermal and its temperature, about 13 °C, is more suitable to its higher latitude. Summer conditions curve 8, are characterized by a thermostad centered at 13 °C and a well developed seasonal thermocline. The thermostad in this region cannot be reliably placed at 13 °C; in other years it has occurred at temperatures above 14 °C and in some years is virtually absent.

According to Istoshin (1961) and Worthington (1959) 18 °C water is found throughout the entire Sargasso Sea. They have concluded that it is formed south of the Gulf Stream and that excess quantities of it flow to the south at the 400–300 m level. It can be detected as far south as the 20th parallel.

Masuzawa (1969) has termed the similar Pacific water mass "subtropical mode water" and has shown that it occupies an even larger area.

In order to form water masses such as these in excess quantities it is not sufficient merely to have outbreaks of cold air over the ocean in winter. Unless there is a net annual heat flux from the oceans to the atmosphere the effect of winter will be merely to remove the heat stored during the summer in the seasonal thermocline. According to Budyko (1956) the net annual heat flux from the northwest Pacific is about 40×10^3 cal/cm^2/y. Over the western North Atlantic the annual heat flux is the greatest found anywhere in the oceans ranging from 40×10^3 to more than 100×10^3 cal/cm^2/y. The relatively weak heat flux in the western North Pacific is probably due to the modification of the continental polar air mass as it passes across the Sea of Japan (Manabe 1957).

A hypothetical north–south section representing the formation of 18 °C water in the northwestern Sargasso Sea is shown in Figure 4. The Gulf Stream is situated at 38° N; it is characterized by the steep descent of the main thermocline, here indicated by the 10 °C and 17 °C isotherms, from the slope water in the north to the Sargasso Sea and by the presence of a "warm core" of surface water (relatively fresh) transported by the Gulf Stream from lower latitudes. This warm core is a reliable feature of the Gulf Stream in the northwestern Sargasso Sea but as it flows toward the east in winter it gradually releases its heat to the atmosphere and it is no longer present south of the Grand Banks where surface Gulf Stream water is indistinguishable from Sargasso Sea water both in temperature and salinity (Fuglister, 1963; Mann 1967). The Gulf Stream can presumably be eliminated as a source of 18 °C water in the northwestern Sargasso Sea but cannot be so eliminated further to the east.

The solid isotherms shown in Figure 4 are drawn roughly at the depths at which they occur in the ocean. South of the Gulf Stream the main thermocline rises gradually toward the south. The current arrows in the upper portion of Figure 4 represent the north/south water movement occasioned by the formation of 18 °C water. A region of convergence is situated immediately to the south of the Gulf Stream. Converged water flows to the south along the core-layer (Wüst 1935) of the 18 °C water with gradually diminishing strength. Surface water is drawn from the south to replace it. This surface water loses heat to the atmosphere in increasing amounts as it moves north (indicated by the wavy arrows at the sea-surface) until its temperature is reduced to about 18 °C at which point it sinks to the 400 m level and flows off to the south.

The lower portion of Figure 4 shows the dynamic height of the 200 m surface relative to the 4000 m surface from two oceanographic sections made

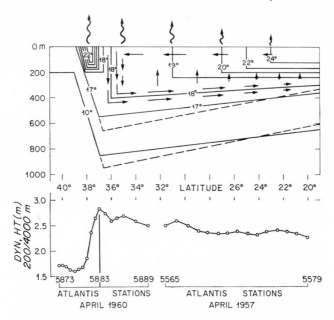

Figure 4   Schematic circulation of 18° water in the Sargasso Sea

from Atlantis. (The northern section Stas. 5873–5889 is from Fuglister, 1963, and runs from the continental shelf to the 33rd parallel N along 68° 30′ W. The southern section, Stas. 5565–5579, runs from Bermuda in a southeasterly direction to 20° N, 55° W.) It is introduced here to illustrate the east/west movement in the Sargasso Sea. The Sargasso Sea can be regarded as an anti-cyclone with its high center situated immediately south of the Gulf Stream. In Fuglister's (1963) Atlantis section the center of the high is at station 5883. North of this high the Gulf Stream flows rapidly toward the east and south of it returning Gulf Stream water flows slowly toward the west over a wide range of latitude. The high occurs at the greatest depth of the thermocline in the convergence region where 18 °C water is most abundant.

The depths of the thermocline isotherms 10 °C and 17 °C, as illustrated here by dashed lines, are essential to the hypothesis which is the basis of this paper. That is, that 18 °C water is formed in excess quantities at the end of the winter and that this results in the deepening of the main thermocline south of the Gulf Stream. If this hypothesis is correct the winter outbreaks of continental polar air are at least in part responsible for forming and maintaining anticyclonic circulation in the Sargasso Sea. Again it must be emphasized that this hypothesis depends on a mean annual heat flux from the ocean to the atmosphere south of the Gulf Stream. It can be noted in Figure 4 that the 10 °C isotherm, a mid-thermocline isotherm, is shown to

The Shape of the Warm Surface Layer
in a Subtropical Gyre

RICHARD P. SHAW

Environmental Science Services Administration, Hawaii Inst. of Geophysics
University of Hawaii, Honolulu, Hawaii

KLAUS WYRTKI

Department of Oceanography, University of Hawaii
Honolulu, Hawaii, 96822

Abstract   The density structure associated with a subtropical anti-cyclonic gyre, in which a Stommel-type horizontal circulation takes place, is investigated using a two-layer model. With parameters appropriate to the Atlantic Ocean, the solution agrees remarkably well with the boundary between the warm and the cold water sphere as outlined by Wüst. The distance of the deepest point of the warm water sphere from the western boundary is a measure of the frictional dissipation parameter. The square of the maximum depth of the warm water sphere determines the horizontal transport of the gyre; these values are in agreement with observations. Moreover, the solution shows that the cold water surfaces in the northwest corner of the ocean and forms a strong cold wall which separates from the western boundary. A comparison with Charney's inertial boundary layer model for the Gulf Stream demonstrates that the shape of the boundary layer solution is essentially the same as that for the frictional model used here. Generation of a mass transport consistent with the assumed density structure and driven exclusively by winds would require a wind-stress amplitude of at least twice the observed value. It is consequently argued that other driving mechanisms must be accounted for, such as the continuous addition of thermal energy and the continuous formation of the warm water mass.

INTRODUCTION

The subtropical anticyclonic gyres are the major circulation systems in the oceans, except for the Antarctic Circumpolar Current. These gyres situated in each hemisphere of each ocean are considered to be essentially wind-driven by the trade winds and the westerlies. Stommel (1948) and Munk

(1950) have derived models of such wind-driven gyres, assuming stationary conditions, omitting the inertia terms, allowing a linear variation of the Coriolis parameters with latitude, and balancing the wind stress by friction. Integrating the equations of motion from the sea surface to the bottom allows the introduction of a two-dimensional horizontal stream function, and forming the vorticity equation eliminates the pressure terms completely. Consequently it is argued that the distribution of density is of no consequence to the circulation derived. The solutions exhibit a remarkable westward intensification of the gyres, and a very strong similarity to the observed pattern of surface currents. However, the computed mass transports of the circulation fall short of the observed transports by a factor of $\frac{1}{2}$, as stated by Munk (1950).

Stommel (1965) tried to overcome this discrepancy by forcing a thermohaline circulation onto this model, achieving slightly greater transports in the western boundary currents.

Very little regard has, however, been given to the fact that the subtropical anticyclonic gyres are characterized by a very pronounced temperature and density structure, as can be seen from the maps published in the *Meteor* Atlas (Wüst and Defant, 1936). Since the circulation derived in both Stommel's and Munk's models is essentially geostrophic, there must be a strong relation between the density structure and the circulation in these gyres. This relation will be investigated in this article. The warm water sphere, as outlined by Wüst (1949) and contrasted to the cold water sphere, is that large water mass in which the circulation of the subtropical gyres takes place.

We will attempt to demonstrate that in the presence of an anticyclonic circulation the warm water sphere must assume a certain shape, which may be calculated from a relatively simple model, and which will be compared with the observed shape. We will show that the circulation forces the boundary between the warm water sphere and the cold water sphere to surface before reaching the northern and northwestern boundary, thus causing a separation of the western boundary current from the shore. Finally, we will compare the shape of the boundary between the warm water sphere and the cold water sphere as derived from a linear theory with that derived from a nonlinear inertial boundary layer theory given by Charney (1955) for the region in which the Gulf Stream forms.

FORMULATION OF THE PROBLEM

A two-layer ocean model is assumed with the boundary between layers of density ϱ_1 and ϱ_2 at the depth D, representing the main thermocline. Below the thermocline, all velocities are assumed to be zero. Thus, the entire circulation is confined to the upper layer of density ϱ_1. The motion is assumed to be planar, but the meridional variation of the Coriolis force is kept by allowing the Coriolis parameter f to be a linear function of the north–south coordinate y, $f = f_0 + \beta y$. The boundaries of the ocean are assumed to form a rectangle with the y-axis northwards from $-b/2$ to $+b/2$ and the x-axis eastward from 0 to L. The elevation h of the sea surface relative to an undisturbed ocean is small compared to D. Since the velocities u and v are assumed to be independent of depth in the upper layer, an integrated transport vector V may be introduced with the components

$$V^x = [(D + h) u] = \int_{-D}^{h} u \, dz \quad \text{and} \quad V^y = [(D + h) v] = \int_{-D}^{h} v \, dz$$

which can be made to satisfy the vertically integrated equation of continuity

$$\frac{\partial}{\partial x} [(D + h) u] + \frac{\partial}{\partial y} [(D + h) v] = 0 \tag{1}$$

automatically through the introduction of a stream function

$$V^x = \frac{\partial \phi}{\partial y}, \quad V^y = -\frac{\partial \phi}{\partial x} \tag{2}$$

The governing equations of motion including Coriolis force, linear frictional forces, geostrophic forces and an applied wind stress at the sea surface, which is assumed to have only a zonal component, are found by vertical integration of the horizontal equations of motion, taking account of the variable limits of integration, and are

$$-f V^y + r V^x - \tau^x = -\frac{1}{\varrho_1} \frac{\partial P}{\partial x} \tag{3}$$

$$f V^x + r V^y = -\frac{1}{\varrho_1} \frac{\partial P}{\partial y} \tag{4}$$

where r is a frictional coefficient. The vertically integrated pressure P is related to the depth of the upper layer D by the equations

$$\frac{1}{\varrho_1} \frac{\partial P}{\partial x} = g \, (D + h) \frac{\partial h}{\partial x} = g \, (D + h) \frac{\varDelta \varrho}{\varrho} \frac{\partial D}{\partial x} \simeq \frac{g'}{2} \frac{\partial}{\partial x} D^2 \tag{5}$$

$$\frac{1}{\varrho_1} \frac{\partial P}{\partial y} = g \, (D + h) \frac{\partial h}{\partial y} = g \, (D + h) \frac{\varDelta \varrho}{\varrho} \frac{\partial D}{\partial y} \simeq \frac{g'}{2} \frac{\partial}{\partial y} D^2 \tag{6}$$

since $h \ll D$, and using $g' = g \, \varDelta \varrho / \varrho$, where $\varDelta \varrho = \varrho_2 - \varrho_1$ is the density difference between the upper and the lower layers.

13 Gordon I

The motion is assumed to be steady, eliminating the independent variable time. The nonlinear inertial terms are neglected, although their influence will be discussed later. The winds over the ocean basin are basically the trade winds in the equatorial half and the westerlies in the poleward half, and are assumed to have only a zonal component of the form $\tau^x = \tau_0 \sin \pi y/B$, although other forms such as a cubic function of y could be used equally well. Frictional effects, representing the dissipation of the energy added by the wind stress, are included as proportional to the volume flux.

By eliminating P between (3) and (4), the volume transport function $\phi\,(x, y)$ defined by (2) is found to satisfy

$$\frac{\partial f}{\partial y}\frac{\partial \phi}{\partial x} + r\left(\frac{\partial^2 \phi}{\partial x^2} + \frac{\partial^2 \phi}{\partial y^2}\right) = \frac{\pi \tau_0}{B}\cos\frac{\pi y}{B} \tag{7}$$

being the same equation as that found by Stommel (1948), but with a different origin for y. Together with the boundary conditions

$$\phi\,(0, y) = \phi\,(L, y) = \phi\,(x, -B/2) = \phi\,(x, +B/2) = 0$$

the formal solution is

$$\phi\,(x, y) = \frac{B\tau_0}{\pi r}\cos\frac{\pi y}{B}\,[c\,e^{\lambda_1 x} + d\,e^{\lambda_2 x} - 1] \tag{8}$$

where λ_1, λ_2 are the roots to

$$r\lambda^2 + \beta\lambda - \frac{\pi^2 r}{B^2} = 0$$

and

$$c = 1 - d = \frac{1 - e^{\lambda_2 L}}{e^{\lambda_1 L} + e^{\lambda_2 L}}$$

We may now return to the equations (5) and (6) and consider the variable depth $D\,(x, y)$ of the boundary between the upper and the lower layer as the dependent variable of interest. These expressions for D^2 together with the solution (8) of equation (7) may be substituted into equations (3) and (4), which are then integrated to find D^2. Calling $D\,(0, 0) = D_0$, the integration gives

$$\frac{g'}{2}\,(D^2(x, y) - D_0^2) = \frac{\tau_0 B}{\pi r}\left\{\frac{\beta B}{\pi}\sin\frac{\pi y}{B} - [f_0 + \beta y]\cos\frac{\pi y}{B}\right\} \times$$

$$\times \{c\,e^{\lambda_1 x} + d\,e^{\lambda_2 x} - 1\}$$

$$+ \frac{\tau_0 B^2}{\pi^2}\sin\frac{\pi y}{B}\,[\lambda_1 c\,e^{\lambda_1 x} + \lambda_2 d\,e^{\lambda_2 x}] \tag{9}$$

We may note at this point, that along $y = 0$ the depth of the upper layer D and the volume transport ϕ are directly related

$$\frac{g'}{2} [D^2(x, 0) - D_0^2] = \frac{B\tau_0}{\pi r} (-f_0) [c\, e^{\lambda_1 x} + d\, e^{\lambda_2 x} - 1] = -f_0 \phi (x, 0) \quad (10)$$

Thus $D(x, 0)$ begins from $D(0, 0)$ at $x = 0$, reaches its maximum value when $\phi(x, 0)$ is maximum and equals $D(0, 0)$ again at $x = L$, where ϕ is zero. This relationship was noticed by Charney (1955) and also by Robinson (1963, editorial note), but given only for the western inertial boundary layer which was the region of interest in that paper. However, that two-layer model neglected friction, but kept the nonlinear terms. Clearly, this result is independent of the forces, other than geostrophic and Coriolis, included in the fundamental equations, and represents a vertical integration of the down-stream equation of motion along the x-axis assuming only a geostrophic balance.

NUMERICAL VALUES USED IN THE MODEL

Although certain physical restrictions have been implied by the form of the equations of motion used in this treatment, thereafter the only assumption made in the derivation of equations (7) and (9) for ϕ and D respectively was that $h \ll D$, i.e., that $\Delta\varrho/\varrho \ll 1$. Next it is of interest to find some approximate forms of these solutions consistent with physically realistic parameters which may yield a deeper insight into the parametric dependence of ϕ and D. We shall consider the North Atlantic as an example.

Using a basin extending from 15° N to 45° N and 80° W to 10° W, we have

$$B = 3330 \text{ km} \simeq 3.4 \times 10^8 \text{ cm}$$

$$L = 7770 \text{ km} \simeq 7.8 \times 10^8 \text{ cm}$$

with an origin at 30° N, where

$$f_0 = 0.730 \times 10^{-4} \text{ sec}^{-1}$$

$$\beta = 0.198 \times 10^{-12} \text{ cm}^{-1}\text{sec}^{-1}.$$

The determination of the density difference $\Delta\varrho$ between the upper and the lower layers is very important insofar as it in turn determines the depth of the density surface with which the model has to be compared. In fact, the depth D of the upper layer and the density difference $\Delta\varrho = \varrho_2 - \varrho_1$ between the two layers are related to the actually observed vertical density distribution $\varrho(z)$ by

$$D\, \Delta\varrho = \int_0^\infty [\varrho_2 - \varrho(z)]\, dz \quad (11)$$

The vertical density distributions near Bermuda, as representative for the center of the subtropical gyre, and south of Greenland, representing the cold water sphere or the lower layer, are shown in Figure 1, as well as the density difference. The data were taken from the *Meteor* Atlas (Wüst and Defant, 1936) and from the Oceanographic Atlas of the North Atlantic Ocean (U.S. Naval Oceanographic Office, 1967). Choosing a value $\Delta\varrho = 1.5 \times 10^{-3}$ gives a maximum depth of the upper layer $D_{\text{max}} = 850$ m near Bermuda according to equation (11). The temperature at that depth is about 12°C, and consequently our results should be compared with the topography of the 12° isotherm. If a density difference $\Delta\varrho = 1.25 \times 10^{-3}$ had been used, the comparison would have had to be made with the 9° isotherm, or with the 16° isotherm if $\Delta\varrho = 2.0 \times 10^{-3}$ had been used.

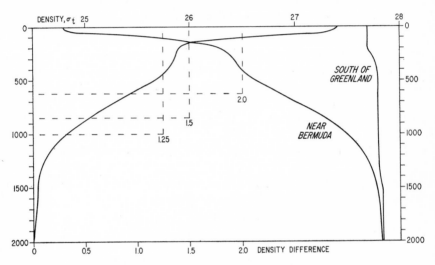

Figure 1 Vertical distributions of density near Bermuda and south of Greenland, together with the density difference between the two locations and equivalent density differences and maximum depths of the upper layer in a two-layer model

To numerically evaluate equations (8) and (9), the wind-stress amplitude τ_0, the frictional coefficient r, and the depth of the upper layer at the latitude of zero wind stress at the western boundary $D(0, 0)$ have to be known. All these three parameters are not easily measured, and some uncertainty exists with regard to their numerical values. Consequently three other parameters shall be introduced in their place. These are the maximum depth of the upper layer D_{max}, the distance from the western boundary where this occurs X_{max}, and the depth of the upper layer at the eastern side of the ocean D_L. These quantities appear to be much more stable—as regards oceanographically measured values—than the wind stress, the friction coefficient and the depth of the thermocline in a region of a very sharp gradient.

It might be immediately pointed out that at $y = 0$ according to (10) the depth of the upper layer at the eastern boundary D_L is precisely the same as that at the western boundary, since $\phi = 0$ in both locations, but in the east this depth changes only slowly with longitude, while in the west the thermocline slopes rapidly upwards and even separates from the western boundary, as will be shown later. From equation (9) it can be learned that the depth of the upper layer along the eastern boundary should be essentially constant within the range of values used here. This is only approximately correct for the real ocean, because upwelling processes along the eastern boundary, which are not included in our model, distort the thermal structure to a certain extent. An inspection of the *Meteor* Atlas reveals that between 20° N and 45° N temperature at 400 m depth varies only between 11° and 13°C within several hundred kilometers off the coast, in a complete contrast to the conditions along the western side of the ocean. Consequently the choice of the depth of the thermocline along the eastern boundary of the ocean as basic input parameter of the model seems to be well justified.

Although wind-stress amplitude, τ_0, and friction coefficient, r, are not particularly well known, they may be taken to be of order 1 dyne cm^{-2} and 10^{-6} sec^{-1}, respectively. With these values, the total volume transport will be taken on the order of 10^{14} cm^3 sec^{-1}.

Using these orders of magnitude for the parameters of the problem, λ_1 and λ_2 may be found approximately from:

$$\lambda_1 \simeq \frac{r}{\beta}\frac{\pi^2}{B^2}, \quad \lambda_2 \simeq -\frac{\beta}{r}$$

Note that $|\lambda_1| \ll |\lambda_2|$ for the physically relevant range of values. We may also find approximate equations for c and d

$$c \simeq e^{-\lambda_1 L}$$

and

$$d \simeq 1 - e^{-\lambda_1 L}$$

The maximum value of D, D_{max}, occurs at $x = X_{max}$ and $y = 0$, where X_{max} is defined by

$$\lambda_1 c\, e^{\lambda_1 X_{max}} + \lambda_2 d\, e^{\lambda_2 X_{max}} = 0$$

which gives the approximate expression

$$X_{max} \simeq \frac{r}{\beta}\ln\left\{\frac{\beta L}{r} + \frac{1}{2}\frac{\pi^2 L^2}{B^2}\right\} \tag{12}$$

This relationship clearly shows that the distance of the deepest point of the thermocline from the western shore is determined by the friction coefficient, and directly proportional to it. The relation (12) holds within a few percent for the range of values in question. Because this distance is a parameter well known from oceanographic observations, it can be used to determine the friction coefficient. The deepest point of most of the isothermal layers in question is found southwest of Bermuda, approximately 525 km from the coastline, which together with other dimensions of the Atlantic Ocean results in a frictional coefficient $r = 2.3 \times 10^{-6} \ \text{sec}^{-1}$. This number is compatible with friction coefficients quoted elsewhere in literature.

From equation (10) the maximum transport ϕ_{max} may be related to D_{max} by

$$\phi_{max} = -\frac{g'}{2f_0} [D_{max}^2 - D_0^2] \tag{13}$$

D_{max} is considerably larger than D_0, which in turn is identical to D_L, the depth of the thermocline at the eastern coast. The total transport of the circulation is determined by these two values. For example, with $g' = 1.5$ and taking $D_{max} = 900$ m and $D_0 = D_L = 340$ m, ϕ_{max} is determined as $71 \times 10^{12} \ \text{cm}^3 \ \text{sec}^{-1}$ in reasonable agreement with the observed circulation.

The depth along the southern boundary changes relatively slowly and is given approximately by

$$D(x, -B/2) - D^2 = -\frac{2B^2\tau_0}{g'\pi^2} \left\{ \left(\frac{\beta}{r} + \frac{r\pi^2}{\beta B^2} \right) c \, e^{(r\pi^2 x/\beta B^2)} - \frac{\beta}{r} \right\}$$

$$= \frac{2B^2\tau_0}{g'\pi^2} \cdot \frac{\beta}{r} \left\{ 1 - e^{-(r\pi^2/\beta B^2)(L-x)} \right\} \tag{14}$$

At the southwestern corner, $x = 0$, $y = -B/2$, the depth is

$$D(0, -B/2) - D_0^2 = \frac{2B^2\tau_0}{g'\pi^2} \frac{\beta}{r} \left\{ 1 - e^{-(r\pi^2 L/\beta B^2)} \right\}$$

$$\simeq \frac{2\tau_0 L}{g'} \left\{ 1 - \frac{1}{2} \frac{r\pi^2 L}{\beta B^2} \right\}$$

while at the southeastern corner, $x = L$, $y = -B/2$,

$$D(L, -B/2) \simeq D_0$$

Finally an expression for ϕ_{max} may be found from (8), which can be approximated by

$$\phi_{max} \simeq -\frac{B\tau_0}{\pi r} \left\{ 1 - e^{-(r\pi^2 L/\beta B^2)} \right\} \cdot \left\{ 1 - \frac{r/\beta L}{(1 + \frac{1}{2} r\pi^2 L/\beta B^2)} \right\}$$

$$\simeq -\frac{\tau_0 \pi L}{\beta B} \left\{ 1 - \frac{1}{2} \frac{r\pi^2 L}{\beta B^2} \right\} \tag{15}$$

relating ϕ_{max} and the wind-stress amplitude τ_0. It should be noted that the total transport is not directly influenced by the frictional coefficient. Unfortunately this last expression is somewhat less accurate; it has an error of 5 to 10%, being larger than the previous approximations, which give errors of only one or two percent.

These approximate equations are not meant to be used in the computer programs to calculate numerical values for $\phi(x, y)$ and $D(x, y)$, but are useful in examining the relationships between various parameters of the model. For example, the friction coefficient appears to have a direct influence only on the distance of the deepest point in the thermocline from the western coast, X_{max} while in the expressions for ϕ_{max} and D_{max} it contributed only to the second order terms.

Using three oceanographically easily measurable parameters, X_{max}, D_{max} and D_L, one can in turn determine those parameters, which need to be known to evaluate numerically equations (8) and (9), namely r, τ_0 and D_0, which are much less accurately observed. From the approximate values of $\phi_{max} = 71.2 \cdot 10^{12}$ and $r = 2.3 \cdot 10^{-6}$, the wind-stress amplitude can be calculated according to (15) and results in $\tau_0 \simeq 3$ dynes cm^{-2}. This value of τ_0 is larger than that usually accepted as the measured value of wind stress. Unfortunately, the transports calculated with the lower, measured value of τ_0 generally appear to fall far below the measured transport values, as observed by Munk (1950). The volume transport function according to Stommel's solution, equation (8), with the above values of r and τ_0, giving a total transport of $71.2 \cdot 10^{12}$ cm^3 sec^{-1}, is shown in Figure 2.

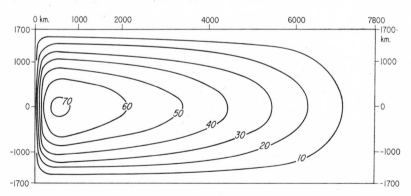

Figure 2 Volume transport function according to Stommel's solution, equation (8), with a total transport of $71.2 \cdot 10^{12}$ cm^3 sec^{-1}

THE SHAPE OF THE WARM UPPER LAYER

With the numerical values derived in the previous section, i.e., $\tau_0 = 3.0$ dynes cm^{-2}, $r = 2.3 \cdot 10^{-6}$ sec^{-1} and $D_0 = 340$ m, a solution of equation (9) has been obtained, and is shown in Figure 3. The choice of $g' = 1.5$ cm sec^{-2} makes this solution comparable to the depth of the 12° isotherm, representing roughly the shape of the warm water sphere in the range of the subtropical anticyclonic gyre of the North Atlantic Ocean, Figure 4. A comparison of the two figures demonstrates the striking similarities between the model and the observations. Along the eastern boundary the depth of the warm upper layer changes little, and on the southern boundary it increases slowly from east to west. The most striking feature of the solution is, however, the surfacing of the cold lower layer in the northwestern portion of the basin and along its northern border. This surfacing is forced by the circulation in the warm layer. South of 30° N the westward flow requires a relatively modest slope of the thermocline, while north of 30° N the corresponding transport to the east requires a much larger slope of the thermocline because of the larger Coriolis parameter. If the total volume of the warm water sphere is limited, this stronger slope forces the cold water to surface. With the parameters used to calculate the shape shown in Figure 3, separation of the warm surface layer from the western boundary occurs 620 km to the north of the latitude of the maximum wind-stress curl.

If the total volume of the warm water sphere is decreased, so that $D_{max} = 840$ m, and the same strength of the circulation is maintained, the shape of the thermocline becomes more pronounced, Figure 5. Along the eastern boundary the cold water comes much closer to the surface, to about

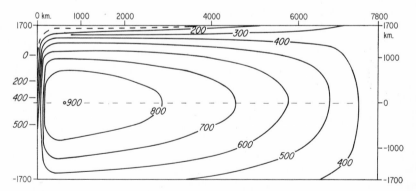

Figure 3 Thickness, in meters, of the warm upper layer of a subtropical anticyclonic gyre according to equation (9). Parameters used: $g' = 1.5$ cm sec^{-2}, $X_{max} = 525$ km, $D_{max} = 900$ m, $D_0 = D_L = 340$ m, corresponding to $\phi_{max} = 71.2 \times 10^{12}$ cm^3 sec^{-1} and $\tau_0 = 3$ dynes cm^{-2}

cates that the two mechanisms of friction and inertia must have very similar effects on the flow pattern even though the processes and mathematical descriptions are quite different.

SUMMARY AND CONCLUSIONS

A Stommel-type solution for the circulation of a subtropical anticyclonic gyre is forced entirely into the warm upper layer of a two-layer ocean and the corresponding shape of this layer is calculated. The calculated shape agrees remarkably well with the observed shape of the warm water sphere as outlined by Wüst (1949). The resulting, chiefly geostrophic adjustment of the thermocline forces the cold lower layer to surface along the western and northern boundaries, causing the separation of the western boundary current from the western shore. The location of this separation point depends critically on the total transport and on the total volume of warm water.

The total transport in the gyre, which is in agreement with the observed shape, is about $70 \cdot 10^{12} \, cm^3 \, sec^{-1}$, and would require a wind-stress amplitude of 3 dynes cm^{-2}, a value much larger than that commonly derived from climatological wind data. Consequently an additional driving mechanism should be in operation, and it might be argued that this driving mechanism is given by the continuous addition of thermal energy into the warm upper layer. After separating from the western shoreline, the strong western boundary current will mix intensively with the water of the cold lower layer, now being beside it. This mixing is indicated in our model by the overspilling of a considerable amount of water across the surfacing line. Consequently the warm water sphere continuously loses warm water and gains cold water, and is thus being slowly eroded. Only the constant formation of the warm water mass by addition of thermal energy can maintain stationary conditions and a constant volume of warm water. It might very well be possible that this addition of thermal energy also constitutes a driving mechanism for the horizontal circulation in the warm water sphere.

Another point of great interest is the fact that the presented frictional solution gives essentially the same transport distribution and the same shape of the upper layer as the inertial boundary-layer solution given by Charney (1955) in the southwestern part of the region, where his solution is applicable. This indicates that the two mechanisms of inertia and friction must have rather similar effects on the large-scale features of the structure and circulation near the western boundary despite their different mathematical formulation. Charney's solution is however limited to that part of the ocean where the boundary layer touches the coast, while the frictional solution is applicable to the entire ocean and describes the separation of the western boundary current from the shore, the formation of the cold wall, and the surfacing of the cold lower layer in the northwestern part of the region.

Acknowledgments

This study was supported partly by the National Science Foundation Grant GA-10277 and by the Office of Naval Research Contract N00014-70-0016-0001, and this support is gratefully acknowledged.

Richard P. Shaw was on an NRC-ESSA postdoctoral research associateship at the University of Hawaii while on leave from the State University of New York at Buffalo.

References

Charney, J.G., The Gulf Stream as an inertial boundary layer, *Proc. National Acad. Sci.*, **41**, 10, 731–740, 1955.

Defant, Albert, *Physical Oceanography*, *Vol. I*, 729 pp., Pergamon Press, New York, 1961.

Munk, W.H., On the wind-driven ocean circulation, *J. Met.*, **7**, 2, 79–93, 1950.

Robinson, Allan R., Editor's Notes, *Wind-Driven Ocean Circulation*, Blaisdell Publishing Company, New York, 1963.

Stommel, H., The westward intensification of wind-driven ocean currents, *Trans., Amer. Geophys. Union*, **29**, 2, 202–206, 1948.

Stommel, H., *The Gulf Stream*, Second Edition, 248 pp., Univ. of California Press, Berkeley, 1965.

U.S. Naval Oceanographic Office, *Oceanographic Atlas of the North Atlantic Ocean*, Pub. No. 700, Washington, D.C., 1967.

Wüst, Georg, Über die Zweiteilung der Hydrosphäre, *German Hydrographic Journal*, **2**, 5, 1949.

Wüst, Georg, and Albert Defant, *Atlas zur Schichtung und Zirkulation des Atlantischen Ozeans*, In Deutsche Atlantische Expedition auf dem Forschungs- und Vermessungsschiff *Meteor*, 1925–1927, Bd. 6, Atlas (103 pls), Walter de Gruyter & Co., Berlin, 1936.